WITHDRAWN

NONLINEAR NETWORKS AND SYSTEMS

NONLINEAR NETWORKS AND SYSTEMS

RICHARD CLAY

WILEY-INTERSCIENCE

a Division of John Wiley & Sons, Inc.
New York · London · Sydney · Toronto

Library of Congress Catalogue Card Number: 76-127660

ISBN 0 471 16040 7

Printed in the United States of America

10 9 8 7 6 5 4 3 2 1

PREFACE

Quite a few years ago I was asked to teach some postgraduate engineering courses for the University of Florida. When discussing nonlinear systems it was necessary to refer to widely scattered references because at that time there was no book which covered the desired material. I decided to write such a book. This was necessarily a part-time project and before it was completed several books on nonlinear systems appeared.

This might have been the end of the project except for another event. A quality control inspector who insisted that all components lie on only one side of a printed circuit board asked me to investigate what restrictions this placed on the circuit. This led to a study of graph theory and the ultimate decision to revive the book and to orient it toward a combination of nonlinear networks and systems.

It remained a part-time activity and, as might be anticipated, before it was completed there appeared the book by Stern on nonlinear networks and systems. However, by very good fortune his book and my outline approached all material from a radically different point of view and hence the treatises are complementary rather than competitive.

This book is addressed primarily to the research engineer or industrial scientist rather than the university student and therefore it contains only suggested exercises rather than formal problems. It contains seven chapters each of which treats a separate topic. Although these are unrelated they are placed in what is hopefully a logical sequence and numerous references between chapters provide continuity.

The first chapter summarizes some significant results in graph theory. Due to my suspicion that automatic processing of electronic circuits on flat surfaces will become increasingly popular there is a rather long discussion of planar graphs. A simple proof of the Kuratowski criterion is given along with an elementary treatment of the work of Whitney on dual graphs.

The second chapter is probably the most difficult. Its purpose is to demonstrate and compare various methods for generating a set of equations which describe a network. It briefly reviews the results of classical dynamics and presents some recent developments. Then the laws of Kirchhoff are combined with some fundamental concepts of graph theory to show how many variables and how many equations one may expect. Some new functions and new techniques for generating equations in various forms are also described and compared.

The appearance of the material of the third chapter in a book of this kind may be somewhat astonishing. It is, however, rather central to the entire discussion. Fitting curves to experimentally derived information and determining distortion products are common engineering exercises. And there is almost universal misconception about the sanctity of a Fourier series representation. The methods of the chapter are also useful, for example, in calculating the describing functions which are used later.

The fourth chapter derives the power formulas in a slightly different manner. The method provides a new set of formulas which heretofore have not been published. The chapter also includes some results which may be useful for electronic engineers.

A brief summary of linear system analysis is given in the fifth chapter along with some standard extensions into the nonlinear domain. The sixth chapter discusses the phase plane with particular emphasis on methods which do not require the tedious construction of isoclines. The seventh chapter presents a very elementary introduction to the method of Liapunov. The primary purpose of this chapter is to provide a simple and logical explanation of the method and to prepare the reader for the available literature which is generally quite advanced.

Notably absent are matrices. While the use of matrices has become almost essential in linear analysis I have thought that the limited advantage of their use in describing nonlinear networks and systems is outweighed by the advantage of presenting the material in a form which is palatable to an obviously larger audience.

Nonlinear differential equations are necessarily scattered throughout the book but discussion on finding explicit solutions to special equations has been avoided because the subject is at present rather unsatisfying and, further, I am not qualified to speak with confidence on even its present state.

Many thanks are due to numerous friends and associates whose assistance has consisted not so much of direct contributions to the manuscript as morale-boosting encouragement which has prompted me to continue the project through many years to its completion.

RICHARD CLAY

Dania, Florida
March 1970

CONTENTS

4 THE POWER FORMULAS

5 LINEARIZATION AND STABILITY

6 THE PHASE PLANE

7 THE LIAPUNOV METHOD

**Phase space, equilibrium points, positive definite and semidefinite
functions, Liapunov function**

NONLINEAR NETWORKS AND SYSTEMS

1

GRAPH THEORY

In 1852 the English scholar, Francis Guthrie, disclosed an interesting theorem to his brother's teacher, Augustus de Morgan of the University College of London. The professor was unable to find a proof and forwarded the problem to the eminent Irish mathematician, Sir William Rowan Hamilton. He also was unable to prove it. Today, more than one hundred years later, the problem remains unsolved and it has become probably the most celebrated conjecture in modern mathematics.

The theorem states that any map drawn on a plane surface can be colored in four colors or less. To color a map means that the colors must be selected so that contiguous countries have different colors. The theorem has been proved for all maps with 35 or fewer countries so any exception must be quite complex.

In their efforts to prove the theorem mathematicians have made typical abstractions of maps. In one representation countries are indicated by points and whenever countries have a common boundary a line connects their respective points. The resultant figure is called a *graph*. Therefore a graph may be defined as a set of points and a set of lines which connect some of the points in a prescribed manner. In addition to the map coloring problem graph theory has been stimulated by work in such diverse fields as molecular structure, formal logic, switching circuits, game theory, and logistics. As a consequence it has become a well developed branch of mathematics and several books on the subject are listed as references for this chapter.

A graph is also an abstraction of a network of physical devices. It represents the structure or skeleton of the network. In general the solution of a network problem involves the establishment of numerous mathematical variables representing quantities whose values need to be found and then generating a set of equations whose solutions evaluate the variables. The second chapter deals almost entirely with methods for deriving these equations. It is shown there that the development of the equations depends on (a) the structure of

1

the network and (b) the properties of the devices which it contains. Thus graph theory deals explicitly with one half of the network problem.

If the network contains nonlinear elements then the equations are usually more difficult to solve. The general results of graph theory which are developed in this chapter are extended in the next chapter to produce equations in the simplest form. Graph theory is therefore particularly valuable in the analysis of nonlinear networks.

Due to the variety of activities which have contributed to graph theory there is a diversity of nomenclature. This chapter summarizes some developments by mathematicians and the corresponding terminology is used. The points of the graph are called *vertices* and the connecting lines are called *edges*. In the next chapter these are called *nodes* and *branches* respectively because the material corresponds more closely to engineering literature.

Figure 1.1 shows a typical graph. Since there is an edge or a series of edges joining each pair of vertices the graph is said to be *connected*. This type of graph is considered most frequently. A series of edges wherein consecutive edges have a common endpoint is called a *sequence*. The series 12345637 is a sequence. It should be observed that one of the edges appears twice. A closed sequence is called a *cycle*. An example is the sequence 123456379. A sequence wherein no edge appears twice is called a *path*. The sequence 123456 is a path. If a path traverses no vertex more than once it is called an *arc*. An example is the path 123458. A closed arc is called a *circuit*. These are of particular importance. The arc 1234589 is a circuit. There are six circuits in the illustration and it may be instructive to find all of them. It should be observed that this definition of a circuit is more specific than common engineering usage. In the next chapter a circuit is called a *mesh* to avoid such confusion.

There are some generalizations which should be noted. As shown in part (a) of Figure 1.2 it is possible for several edges to join the same pair of

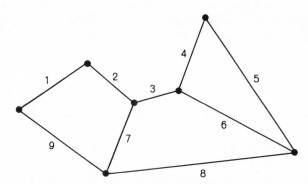

Figure 1.1   A simple graph.

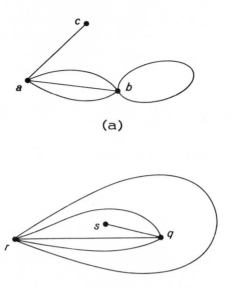

(a)

(b)

Figure 1.2   (a) Some generalizations. (b) An isomorphic graph.

vertices, in this case *a* and *b*. These are called *multiple edges*. If no other edges join vertex *c* then the edge shown between vertices *a* and *c* is called a *pendant edge*. The single edge having both endpoints at vertex *b* is called a *loop*. Pendant edges and loops are primarily of interest to mathematicians since they have little function in a network of physical devices.

If between two graphs there exists a vertex-to-vertex correspondence such that their edges make identical connections then the graphs are said to be *isomorphic*. Part (b) of Figure 1.2 shows a graph which is isomorphic to the graph in part (a). The vertices *q*, *r*, and *s* correspond respectively to vertices *a*, *b*, and *c*.

If edges are joined only to the endpoints of an arc then it is called a *chain*. If two graphs are isomorphic when chains are replaced with single edges then they are said to be *conformal*. Since a chain of edges and a single edge are geometrically equivalent it is apparent that the property of conformality is important in the determination of the geometric characteristics of graphs. It is used later in the discussion of planar graphs.

A graph *g* is called a *subgraph* of the graph *G* if it contains only edges and vertices found in *G*. A subgraph is said to *span* the graph *G* if at least one edge of *g* joins every vertex of *G*. When the edges of *g* are deleted from *G* the result is called the *complement* of *g*.

CONNECTIVITY

There are some very general results in graph theory which depend only on the fact that some points are connected with lines. A few additional definitions are needed. The numbers of edges and vertices in the graph G are designated $e(G)$ and $v(G)$ respectively. If G is composed of several connected parts which are not joined to each other then the number of parts is expressed as $p(G)$. The number of edges joining vertex a is called the *vertex valence* of a and it is designated $\rho(a)$. These definitions immediately lead to the relation

$$e(G) = \tfrac{1}{2} \sum_a \rho(a) \qquad (1.1)$$

since each edge is counted twice in forming the sum. If loops are present they are counted twice at their vertices. The sum remains an even number when all vertices with even valence are deleted and the following theorem is immediately established.

Theorem 1.1 *A graph contains an even number of vertices with odd valence.*

If the deletion of a subgraph g from a connected graph G forms a complement $G - g$ which has several connected parts then

$$v(g) \geqslant p(G - g) \qquad (1.2)$$

since each vertex in g can be joined to no more than one part of $G - g$. This is stated in the following theorem.

Theorem 1.2 *The number of vertices in a subgraph is no less than the number of connected parts in its complement.*

To prove the next theorem it is necessary to introduce another concept. If each vertex in a graph is connected to every other vertex then it is called a *complete graph*. In such a graph it is always true that

$$e(G) = \tfrac{1}{2}v(v - 1) \qquad (1.3)$$

since each of the v vertices is joined to the other $v - 1$ vertices.

Theorem 1.3 *The maximal number of edges in a graph having neither multiple edges nor loops is given by $e_m = \tfrac{1}{2}(v - p)(v - p + 1)$.*

It is quite apparent that for a given number of vertices the complete graph has the most edges. Therefore each connected part of the graph must be a complete graph. Also, given any set of complete graphs it is easily seen that when one transfers vertices from the smaller graphs to the larger ones the requirement for edges is increased. In the given case the maximum occurs when there is one complete graph with $v - p + 1$ vertices plus $p - 1$ isolated vertices to provide a total of p parts. Equation 1.3 then gives the number of edges stated in the theorem.

Corollary 1.3.1 *A graph is connected if* $e(G) > \frac{1}{2}(v - 2)(v - 1)$.

Substituting $p = 2$ in the formula stated in Theorem 1.3 gives the maximal number of edges if the graph remains in two parts. Any additional edges must connect the parts since multiple edges and loops are excluded.

Theorem 1.4 *If the vertices of two graphs can be placed into a correspondence such that the valences are equal at each vertex then one graph can be transformed into the other by a series of edge transfers.*

An edge transfer is an operation where edges (a, q) and (b, r) are replaced with edges (a, r) and (b, q). This has no effect on the vertex valences. The vertex set $(a, b, c, \ldots)$ which was previously connected to the vertex set $(q, r, s, \ldots)$ is now connected to $(r, q, s, \ldots)$. The edge transfer has the effect of interchanging two elements of the second set. Given the permutation $(q, r, s, t, \ldots)$ an interchange of r and s followed by an interchange of q and s produces the permutation $(s, q, r, t, \ldots)$. Thus a series of edge transfers can translate a given element to the first position in the permutation leaving the other elements in the original order. A similar operation can place any of the remaining elements in the second position. Continuing the process to its conclusion transforms any given permutation of vertices to any other permutation. This transforms the given graph to any other graph with the same vertex valences.

Theorem 1.5 *In a connected graph there is at least one arc which contains any two given edges.*

Assume that edges (a, b) and (q, r) are given. Since the graph is connected there exists an arc between vertices b and q. If this arc does not pass through vertices a or r then the required arc is the arc (a, b, q, r). However, if arc (b, q) passes through vertex r, for example, then the required arc is the arc (a, b, r, q).

A vertex a is an *articulation point* of a connected graph if and only if the deletion of all edges joining a separates the graph into more than one part. It is also called a *cut vertex*. If a connected graph has an articulation point it is said to be *separable*. Otherwise it is *nonseparable*.

Lemma 1.1 *A vertex a is an articulation point of a connected graph if and only if there exist two vertices such that every arc which connects them passes through a.*

The necessity of the condition is easily established by choosing vertices which lie in different parts of the graph which is formed by deleting the edges joining *a*. All arcs which connect the vertices must necessarily pass through *a*. The sufficiency of the condition depends on the fact that if every arc connecting the vertices passes through *a* then the subgraph $G - a$ is no longer connected and hence *a* is an articulation point. The following important theorem is due to Menger.

> **Theorem 1.6** *Between any two vertices of a nonseparable connected graph there are at least two arcs which have only the given vertices in common.*

This is conveniently proved by induction relative to the number of edges in a selected path. The theorem is applicable if there is only one edge between the vertices because if there is no other arc connecting them then both vertices are articulation points. Figure 1.3 shows the general case. Each part of the figure shows the two given arcs between vertices *a* and *b* plus the additional edge (*b*, *c*). Since the graph is connected there must be an arc (*a*, *c*) which is also shown. In part (a) the arc (*a*, *c*) is entirely unique from vertex *a* to vertex *c*. Two disjoint arcs are shown as heavy lines. Another set exists and its discovery is left as an exercise. In part (b) the arc (*a*, *c*) intersects one of the given arcs at vertex *d*. The required disjoint arcs are shown. A few additional cases are possible. If vertex *c* lies inside the two given arcs then the same construction applies. If vertex *c* lies on one of the given arcs then the required

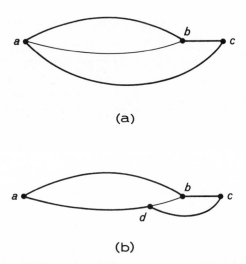

(a)

(b)

Figure 1.3 The disjoint arcs of Menger's theorem.

arcs have the relationship shown in part (b) except that vertices c and d are coalesced to a single point at the location of vertex d.

Corollary 1.6.1   *Any two vertices in a connected nonseparable graph lie on a circuit.*

This follows immediately from Menger's theorem and requires in addition only the definition of a circuit.

Corollary 1.6.2   *In a connected nonseparable graph there exists at least one cycle passing through any three given vertices.*

According to the previous corollary there exist circuits through each pair of vertices. In each circuit at least one arc does not pass through the third vertex since the circuit would then become a path. The sum of three such arcs forms a cycle through the three given vertices. It should be observed that

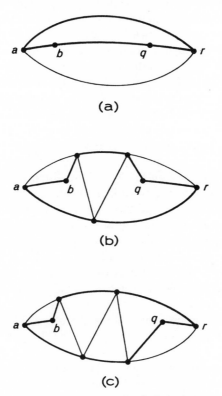

(a)

(b)

(c)

Figure 1.4   Finding a circuit through two given edges.

this cycle is not necessarily a circuit since any pair of the selected arcs may contain a common edge.

Theorem 1.7 *Any two edges in a connected nonseparable graph lie on a circuit.*

Assume that edges (a, b) and (q, r) are given. Since the graph is connected there is an arc between vertices b and q. Since it is also nonseparable there are two arcs between vertices a and r. Figure 1.4 shows the possible relationships which these paths may have. The required circuit is shown as the heavy line in each case. The same construction is used when the two arcs between vertices a and r lie on the same side of the arc (b, q).

PATH PROBLEMS

In addition to the map coloring problem several related topics have stimulated the development of graph theory. Some of these are concerned with the discovery of paths in the graph which have special properties. A very common and practical problem associated with maps is the determination of the shortest path between two towns. In this case the graph is amplified by associating with each edge a number known as the *distance*. A standard iterative procedure then solves the problem.

To each vertex which is removed by one edge from the initial vertex one assigns a distance using the distance of the connecting edge. Then all other edges joining these vertices are examined to determine whether or not an indirect arc including several edges is shorter than the direct path. If such an arc exists then the vertex distance is reduced accordingly.

One then considers in arbitrary order each remaining vertex which is removed by only one edge from vertices whose distances are already assigned. The minimum distance is computed in a similar manner for each. Then one examines every arc from each of these vertices to determine whether or not the selection of an arc through any of them provides a path to previously considered vertices which is shorter than the path previously used. If such a path exists then the vertex distance is reduced accordingly. It is then necessary to reconsider all arcs from this vertex to determine the effect of changing its distance. Any effect must be propagated throughout the graph.

The process is continued until the distances of all vertices have been reduced to their minimum values. Since there is a finite number of edges these minimum distances form a unique set. The shortest arc to the terminal vertex is the one used to calculate its distance and this is the solution to the problem. Figure 1.5 shows a map where the edge distances have been assigned. It is not drawn accurately to scale. The minimum vertex distances have been calculated for the first set of vertices. It is left as an exercise to show that the minimum distance between vertices a and b is 13.

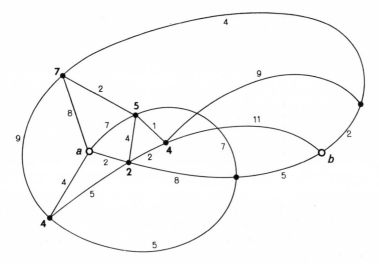

Figure 1.5 Determining the shortest path.

The problem of finding the arc between two vertices that includes a minimum number of edges is solved similarly except that a unit distance is assigned to each edge.

An interesting example of a problem to determine the shortest distance is the solution of a labyrinth or maze. The methods of graph theory can be used. Each junction in the labyrinth is a vertex in its graph. The lengths of the actual paths in the maze are the edge distances in the graph. The above method then indicates the best route through the maze.

There is an interesting method for solving this and related problems using a string model of the graph. Lengths of string are cut so that they equal the edge distances when tied at the vertices. The model is then grasped at the initial and terminal vertices and gently pulled apart until a series of edges becomes taut. This is the path of minimum distance.

There is another problem related to maps which is quite interesting and practical. It concerns the discovery of a route which passes only once through each town of a prescribed set of towns. The analog in graph theory is the discovery of a circuit which passes once through every vertex. Such a path is called a *Hamilton circuit*. If one edge is deleted it still connects every vertex and it is called a *Hamilton arc*. Figure 1.6 shows a Hamilton circuit in a graph.

Although the concept of a Hamilton circuit is quite simple there is no definitive criterion for determining whether or not such a circuit can be found in a given graph. The following results give some sufficient conditions. But first it is necessary to introduce another concept.

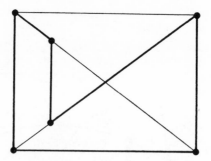

Figure 1.6 A Hamilton circuit.

An arc can be used to define a subgraph. It traverses a unique set of vertices. The set of edges can be defined to include only those edges which connect members of the vertex set. This technique is used to establish the following theorem.

> **Theorem 1.8** *A connected graph which has neither multiple edges nor loops has a Hamilton circuit if the sum of the valences at any two vertices is never less than the number of vertices.*

It is assumed that in graph G the arc (a, b) defines a subgraph g which has $v(g)$ vertices. As shown in Figure 1.7 a circuit can be formed through the same vertices if there exist edges (a, r) and (q, s). Since the graph G has neither multiple edges nor loops such edges must exist if

$$\rho(b) > [v(g) - 1] - \rho(a) \tag{1.4}$$

because the expression on the right gives the number of vertices not joined by edges from vertex a. If the arc can be extended to pass through every vertex in the graph G then the criterion for G to possess a Hamilton circuit becomes

$$\rho(a) + \rho(b) \geqslant v(G) \tag{1.5}$$

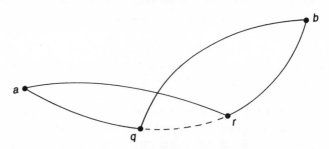

Figure 1.7 Construction of a Hamilton circuit.

Since no pair of vertices occupies a unique position in a circuit the relation must be met by all pairs of vertices in G.

A cube has a Hamilton circuit yet the valence of each vertex is 3 and there are 8 vertices. These numbers do not satisfy Equation 1.5 so the requirement of Theorem 1.8 is sufficient but it certainly is not necessary. The following corollaries are derived directly from the theorem.

Corollary 1.8.1 *A connected graph has a Hamilton arc if* $\rho(a) + \rho(b) \geqslant v(G) - 1$ *for all a and b.*

Corollary 1.8.2 *A connected graph has a Hamilton circuit if* $2\rho(a) \geqslant v(G)$ *for every vertex.*

Corollary 1.8.3 *Every complete graph has a Hamilton circuit.*

There is another type of path through a graph which is of interest. Part (a) of Figure 1.8 shows a very common puzzle. The problem is to trace all edges of the graph once and only once without lifting the pencil from the paper. It is impossible. The graph in Figure 1.8(b) can be traced if one starts at vertex a and the path ends at vertex b regardless of the manner in which it is constructed. The graph in Figure 1.8(c) can also be traced and one can start at any vertex. The path returns to the same vertex.

Euler was the first to analyze the requirements for the existence of such paths in a graph. Consequently a path which passes once along every edge

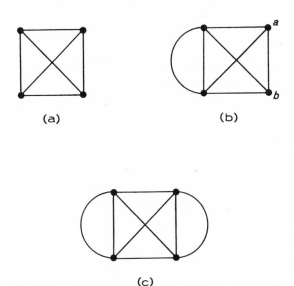

(a)

(b)

(c)

Figure 1.8 Demonstrating the existence of Euler paths.

is called an *Euler path*. If the path returns to the initial vertex it is called a *cyclic Euler path*.

Theorem 1.9   *A connected graph has a cyclic Euler path if and only if every vertex has even valence.*

The condition is obviously necessary since each passage through a vertex involves two edges. To establish the sufficiency of the condition a typical path is constructed. Since each vertex has even valence, whenever a path enters a vertex it is always possible to leave it. Since the graph is connected the path can always return to the initial vertex. However, it may not include all of the edges. In this case one constructs another cyclic path in the remaining graph. This is possible since every vertex again has even valence. If this path does not include all of the remaining edges the process is repeated. Eventually all edges are included in a set of cyclic paths. Since the graph is connected there is no cyclic path which is vertex disjoint from all of the others. Therefore the sum of all cyclic paths produces one cyclic Euler path. The following corollaries state these principles in alternate fashions.

Corollary 1.9.1   *If a graph has a cyclic Euler path then it is the union of edge-disjoint circuits.*

Corollary 1.9.2   *If a graph has a cyclic Euler path then every vertex lies on a circuit.*

Figure 1.8 shows three graphs. One has a cyclic Euler path, one has an ordinary Euler path, and the third requires two paths to include all edges. The following theorem clarifies the situation.

Theorem 1.10   *A connected graph with 2n vertices of odd valence can be spanned with no fewer than n edge-disjoint paths.*

If the $2n$ vertices of odd valence are joined in pairs with n additional edges the result is a graph wherein all vertex valences are even. It contains a cyclic Euler path. This path includes the additional edges. When these edges are removed to produce the original graph the cyclic Euler path is broken into n sections since the set of added edges is not connected. An example of this process is the transition from part (c) to part (a) of Figure 1.8. The removal of the two edges creates two edge-disjoint paths.

TREES

A graph wherein every pair of vertices is connected by exactly one arc is called a *tree*. If a tree is a subgraph which spans every vertex of the entire graph it is called a *maximal tree*. Several trees constitute a *forest*. Figure 1.9

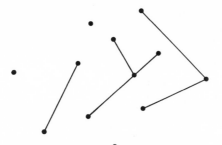

Figure 1.9   A forest of three trees.

shows a forest. The above definition of a tree is not the customary one. It is chosen because it has a position symmetric to the properties of trees which are described in the following lemmas.

Lemma 1.2   *A tree has no circuits.*

Lemma 1.3   *The deletion of any edge of a tree divides it into exactly two parts.*

These lemmas follow directly from the definition. Although they are rather obvious and seemingly unrelated they are actually very significant and constitute the point of departure for a large part of network analysis.

Theorem 1.11   *There are v − 1 edges in a tree.*

This is easily established by disassembling the tree in an orderly fashion. A vertex where $\rho(a) = 1$ is called a *terminal vertex* and the edge which joins it is called a *terminal edge*. One starts at a terminal vertex and removes edges successively until a vertex is reached where the valence is greater than two. The process is then repeated starting each time at one of the remaining terminal vertices. For each edge which is removed there is a corresponding vertex except at the last edge where there are two. The number of edges in the tree is therefore one less than the number of vertices.

Corollary 1.11.1   *There are v − p edges in a forest.*

For each tree in the forest there are $v_i − 1$ edges. Forming the sum over all trees gives v − p edges for the forest. In this formula isolated vertices can be considered as degenerate trees. They increase the number of parts accordingly and the formula remains valid.

Each edge of the complement of a maximal tree is called a *chord*. Figure 1.10 shows a maximal tree and its chords. It is observed that there are eight vertices and seven edges on the tree.

Theorem 1.12   *There are e − v + 1 chords for a maximal tree.*

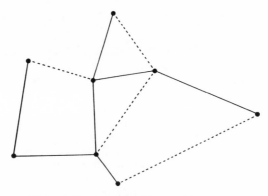

Figure 1.10   A maximal tree and its chords.

Of the e edges which are contained in the graph $v - 1$ edges lie on the tree. Consequently $e - v + 1$ edges are chords.

Corollary 1.12.1   *There are $e - v + p$ chords for a forest.*

This follows immediately by applying the formula of the theorem to each tree in the forest and then forming the sum over all trees. Again isolated vertices can be considered as degenerate trees and the formula remains valid.

   Networks consisting entirely of trees have some very practical applications. They represent, for example, the minimum system of roads which connect a given set of towns or the minimum network to provide communication between given points. Therefore it is sometimes valuable to know how many trees can be constructed on a given set of vertices. The basic problem has an easy solution. The derivation is considerably simplified by arbitrarily enumerating the vertices. The vertex set then becomes

$$V = 1, 2, 3, \ldots, v \tag{1.6}$$

A tree constructed on these vertices can be expressed uniquely by a sequence of $v - 2$ numbers. This is not obvious and is most easily explained by an example. Figure 1.11 shows a tree with enumerated vertices. The terminal vertex set is

$$V_t = 1, 2, 5, 7 \tag{1.7}$$

The first member of this set is connected to vertex 3 so the required sequence starts as

$$S = 3, \ldots. \tag{1.8}$$

The edge $(1, 3)$ is deleted and the new terminal vertex set is

$$V_t' = 2, 5, 7 \tag{1.9}$$

The first member of this set is also connected to vertex 3 so the required sequence continues as

$$S = 3, 3, \ldots \tag{1.10}$$

Then the edge (2, 3) is removed. The next terminal vertex set is

$$V_t'' = 3, 5, 7 \tag{1.11}$$

The process is continued until the tree is reduced to the last edge. This edge joins vertices 6 and 7. However, vertex 7 does not form part of the required sequence. It is the last remaining vertex and it offers no additional permutations. The final descriptive sequence is

$$S = 3, 3, 4, 4, 6 \tag{1.12}$$

This series of numbers corresponds uniquely to the given tree and by reversing the procedure it is possible to construct the original graph from the sequence of numbers. Although there are 7 vertices there are only 5 numbers in the sequence.

In the general case to account for all possible configurations each of the v vertices must appear in each of the $v - 2$ positions. The total number of possibilities is therefore given by the formula

$$N_t = v^{v-2} \tag{1.13}$$

This formula was discovered by Cayley when he was studying molecular structure. It has been used, for example, to predict the number of hydrocarbons which have a single chemical bond between atoms. The formula gives the proper number of trees under the assumption that each vertex maintains its identity. It therefore includes many isomorphic forms.

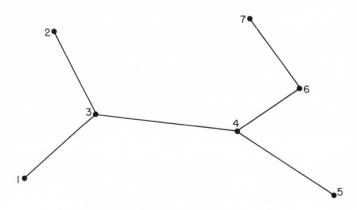

Figure 1.11   A tree with enumerated vertices.

CIRCUITS AND CUT-SETS

A circuit has been defined as a closed arc. Since it passes only once through every vertex along its path a circuit can also be defined as a subgraph wherein every vertex has a valence of 2. The selection of a particular maximal tree permits the definition of a unique set of circuits which are associated with the tree. Figure 1.12 shows a graph and a maximal tree. Its chords are shown as dotted lines. Such a maximal tree can be constructed in an orderly fashion. The edges are enumerated arbitrarily. Then each is examined in order. If it lies in a circuit it is deleted and it forms a chord for that circuit. It is a geometric property of a graph that if the removal of an edge destroys a circuit then it destroys only one circuit. Thus there is a unique relationship. A circuit which corresponds to a chord is called a *basic circuit*. It contains only that chord. All other edges in the basic circuit belong to the tree. The curve C in Figure 1.12 outlines the basic circuit which corresponds to chord (a, b).

The *circuit rank* of any graph is defined by the equation

$$n(G) = e - v + p \tag{1.14}$$

This is immediately recognized as the number of chords in a forest. In each tree of the forest there is exactly one basic circuit for each chord and the following property is established.

Theorem 1.13   *In any graph the circuit rank equals the number of basic circuits.*

The above theorem combined with Lemma 1.2 immediately establishes the following corollary.

Corollary 1.13.1   *A graph is a forest if and only if the circuit rank is zero.*

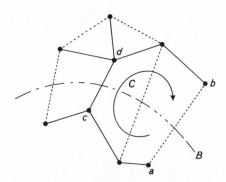

Figure 1.12   Basic circuit and basic cut-set.

It is left as an exercise to show how any circuit in a graph can be made a basic circuit by proper enumeration of the edges in the construction of the maximal tree. This provides the following extension to the theorem.

Corollary 1.13.2   The deletion of an edge reduces the circuit rank by one if and only if the edge lies in a circuit.

The following simple and elegant theorem which is due to Whitney is an interesting characterization of a circuit.

Theorem 1.14   A circuit is a minimal set of edges containing at least one chord of every maximal tree.

Since the set of edges is minimal it contains no edge which is not a chord of a maximal tree. Therefore the deletion of any edge destroys a basic circuit for a particular maximal tree. This reduces the circuit rank by one and according to Corollary 1.13.2 the set of edges is a circuit. This proves the sufficiency of the condition. Assuming that the set of edges is a circuit then the deletion of any edge reduces the circuit rank by one. This means that one basic circuit is destroyed and this indicates the removal of a chord. In addition, if the circuit does not contain a chord for every maximal tree then there exists a tree which creates basic circuits which are not destroyed by removal of the circuit edges. This is contrary to fact and the necessity of the condition is established.

A minimal set of edges whose deletion divides a connected nonseparable graph into two parts is called a *cut-set*. If a finite graph can be represented in a space of 2 or 3 dimensions then it is possible to pass through its edges a surface which divides it into two parts. The edges which the surface intersects form a cut-set and the following theorem is established.

Theorem 1.15   Every connected graph contains a cut-set.

The analogy between circuits and cut-sets is frequently amazing. There are no circuits in a maximal tree. Following is the corresponding lemma for cut-sets.

Lemma 1.4   The chords for a maximal tree do not contain a cut-set.

This is easily established by deleting the entire set of chords. The maximal tree remains and all vertices of the graph are still connected.

Just as a maximal tree is useful in constructing circuits in a graph it can also be used as a basis for cut-sets. The deletion of any edge of the tree divides it into two parts. This partitions the vertices of the graph into two connected parts. A continuous surface can be passed through the deleted tree edge and various chords so as to correspond to the vertex partition. Again there is a unique relationship. The cut-set which corresponds in this manner to a tree edge is called a *basic cut-set*. Curve *B* in Figure 1.12 defines a basic cut-set of edges. In a fashion analogous to a basic circuit the basic cut-set contains one tree edge and all other edges are chords.

The *cut-set rank* of any graph is defined by the equation

$$r(G) = v - p \qquad (1.15)$$

This is immediately recognized as the number of edges in a forest. In each tree of the forest there is exactly one basic cut-set for each tree edge and the following property is established.

> **Theorem 1.16**   *In any graph the cut-set rank equals the number of basic cut-sets.*

In the limiting case where $r(G) = 0$ it is always true that $v = p$ and the graph contains no edges. This leads to the following corollary.

> **Corollary 1.16.1**   *A graph contains only isolated vertices if and only if the cut-set rank is zero.*

By definition a cut-set is the minimal set of edges whose deletion divides the graph into two parts. This applies equally well to any connected part of a graph. In the general case p increases by one while v remains constant so the following corollary is always applicable.

> **Corollary 1.16.2**   *A cut-set is a minimal set of edges whose deletion reduces the cut-set rank by one.*

The following theorem is analogous to Theorem 1.14. Its proof is left as an exercise with the suggestion that the analogy between circuits and cut-sets extends to chords and tree edges and also quite obviously to circuit rank and cut-set rank.

> **Theorem 1.17**   *A cut-set is a minimal set of edges which contains at least one edge of every maximal tree.*

From the theorem it is obvious that the deletion of a cut-set from a graph must make it impossible to construct any maximal tree from the remaining edges and the following corollary is established.

> **Corollary 1.17.1**   *The complement of a cut-set contains no maximal tree.*

The previous theorems in this section involve circuits and cut-sets separately. The following two theorems state some relationships between the two types of sets.

> **Theorem 1.18**   *If a circuit contains edges which belong to a cut-set then the circuit and the cut-set contain an even number of edges in common.*

A cut-set is equivalent to a continuous surface in a graph since every edge which intersects the surface appears in the cut-set. A circuit is a simple closed curve. Such a curve must intersect a continuous surface an even number of times and each intersection is an edge which is common to both sets.

Theorem 1.19 *The basic cut-set for an edge of a tree contains those chords which correspond to basic circuits containing the same edge.*

Except for the given tree edge all other edges in the basic cut-set are chords. According to Theorem 1.18 this cut-set must contain every chord which lies on a circuit passing through the given edge. A basic circuit passes through each chord and only that chord. Since the cut-set contains no tree edges other than the given one the basic circuit must return through the given edge. Thus the chords in the basic cut-set can be placed into a correspondence with the basic circuits passing through the given tree edge. This is illustrated in Figure 1.12 for the edge (c, d).

Theorem 1.20 *The basic circuit for a chord of a tree contains those tree edges which correspond to basic cut-sets containing the same chord.*

This theorem concerning the construction of basic circuits is analogous to the previous one. Its proof is left as an exercise.

DIRECTED GRAPHS

If a sense of direction is assigned to some of the edges of a graph it is called a *directed graph*. Such assignment can arise from two sources. First, there can be a real reason why the flow along an edge can be in only one direction. If, for example, the graph should represent the map of a city then directed edges correspond to one-way streets. When finding the best route between given points it is necessary to account for such restrictions. In the second type, such as the traditional solution of network problems, it is customary to establish arbitrary senses to the edges to provide a reference. There is no physical reason why the actual flow can not be in the opposite direction.

Each vertex may have two valences in a directed graph. The number of edges *leaving* vertex a is called the *positive vertex valence* and is designated $\rho^+(a)$. Conversely the *negative vertex valence* is $\rho^-(a)$ and it is the number of edges which *enter* vertex a. All valences are positive numbers. The total number of edges at vertex a is given by

$$\rho(a) = \rho^+(a) + \rho^-(a) \tag{1.16}$$

Each edge contributes to a positive valence at one vertex and a negative valence at another so over the entire graph it is true that

$$e(G) = \sum_a \rho^+(a) = \sum_a \rho^-(a) \tag{1.17}$$

Normally trees, circuits, and cut-sets in a directed graph are derived from the associated undirected graph where the sense of direction is removed from each edge. However the directed quantities can also be defined. A *directed arc* is an arc which has a sense of direction consistent with all of its edges. A *directed tree* is a tree in the undirected graph which has a directed arc from a special vertex a_0 to all other vertices. The vertex a_0 is called the *root*. Not all directed graphs have a directed tree.

A directed circuit is a directed arc which is closed. It is left as an exercise to show that a directed circuit is a subgraph where $\rho^+(a) = \rho^-(a) = 1$ for every vertex. Not all directed graphs have a directed circuit.

A cut-set in a directed graph has in general three types of edges. They may have either sense of flow across the cut-set surface or they may have no sense of flow whatever. It is sometimes convenient to distinguish between these components but such problems are not considered here. The following theorems and corollaries are trivial for undirected graphs but they have some significance in directed graphs.

Theorem 1.21   *A directed cycle is the sum of directed circuits.*

One can start at any vertex on the cycle and construct a path consistent with the senses of the edges. Whenever the path passes twice through a vertex this indicates a directed circuit. The circuit is deleted. The remaining graph is also a cycle. Another arbitrary directed path is constructed until it forms another circuit. This circuit is also deleted. Continuing the process decomposes the cycle into circuits. Along the last path $\rho^+(a) = \rho^-(a) = 1$ at every vertex so the last path is also a directed circuit.

Corollary 1.21.1   *A directed path is the sum of directed arcs and directed circuits.*

A directed path between two vertices necessarily consists of a collection of directed arcs plus directed cycles. According to the above theorem the latter can be decomposed into directed circuits.

Theorem 1.22   *If the positive vertex valence is greater than zero at every vertex then the graph contains a directed circuit.*

For any initial vertex which may be chosen there is a departing edge since the positive valence is never zero. This edge leads to another vertex where the positive valence again is not zero. The procedure can be continued and since the graph is finite the path ultimately must return to a vertex which has already been passed. This constructs a directed circuit. The theorem is also valid when it involves the negative valences.

A graph is *strongly connected* if for all vertices a and b there exist arcs (a, b) and (b, a).

Theorem 1.23 *If the union of directed circuits is connected it is strongly connected.*

According to Theorem 1.21 the union of the directed circuits forms a directed cycle. This cycle provides paths in both directions between any pair of vertices.

Corollary 1.23.1 *A graph is strongly connected if and only if every cut-set contains at least two edges flowing in opposite senses across the cut-set surface.*

For the graph to be strongly connected it is sufficient that each pair of vertices lies on a directed circuit. Theorem 1.18 implies that any cut-set through such a graph contains at least two edges in common with a directed circuit. These have opposite senses relative to the surface which generates the cut-set.

PLANAR GRAPHS

A graph is *planar* when it has an isomorphism whose vertices lie in a plane and whose edges make contact only at end points. The characteristics of planar figures have always been of interest to mathematicians. Such graphs are acquiring more importance in the practical field because of the increasing use of automatic methods for manufacturing electronic circuits on flat surfaces. To determine criteria for the recognition of such graphs it is necessary to introduce a few geometric properties of plane surfaces.

If a continuous closed curve is drawn in a plane it defines an inner and an outer region. A point in the inner region can not be joined to a point in the outer region without intersecting the curve. This is shown in Figure 1.13(a).

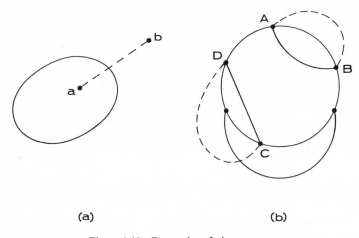

(a)   (b)

Figure 1.13 Properties of planar curves.

Two points along a closed curve can be joined by an *inner transversal* or an *outer transversal*. Frequently a transversal can be transferred from one region to the other by an isomorphic change. This is shown in Figure 1.13(b) as the transversal joining points *A* and *B*. However, if the endpoints of an outer transversal separate those of an inner transversal then neither can be transferred without an intersection. This is shown in Figure 1.13(b) as the inner transversal joining points *C* and *D*.

> **Theorem 1.24**   *If a graph can be disconnected by the removal of two vertices a and b then it is planar if and only if each of the disjoint components plus edge (a, b) form planar graphs.*

Quite obviously the components can be fitted together along edge (*a, b*) and the result is planar if each component plus edge (*a, b*) is planar. However, if any of the components plus edge (*a, b*) form a nonplanar graph then the original graph would be nonplanar since it is connected and hence contains a path in another component which is geometrically equivalent to edge (*a, b*).

In 1930 Kuratowski presented an interesting and important criterion to determine whether or not a geometric figure is planar. His method involves the two graphs shown in Figure 1.14. Each of the figures is shown in two ways.

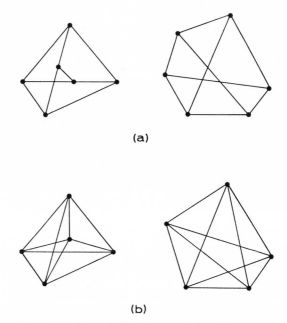

(a)

(b)

Figure 1.14 Isomorphic forms of the Kuratowski graphs. (a) The hexoid and (b) the pentoid.

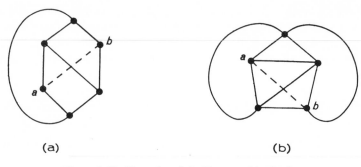

(a)   (b)

Figure 1.15   Necessity of the Kuratowski criterion.

The pictures on the left show pyramids with some internal lines and the pictures on the right show graphs with intersecting edges. It is left as an exercise to demonstrate that the pairs are isomorphic. The first figure is a hexagon with three diagonals. It will be called the *hexoid*. The second figure is a complete pentagon. It will be called the *pentoid*. These definitions permit a concise formulation of the Kuratowski criterion.

Theorem 1.25   *A finite graph is planar if and only if it contains no subgraph conformal to either the hexoid or the pentoid of Kuratowski.*

The necessity of the criterion is easily established. Figure 1.15 shows the figures drawn with all possible inner transversals transferred to the outer region. In each case there remains an inner transversal which forms part of a closed curve such that vertex a is inside the curve and vertex b is outside. The remaining edge cannot be drawn without an intersection.

The proof of the sufficiency of the criterion is more difficult. It must be shown that every nonplanar graph contains a subgraph conformal to at least one of the two Kuratowski graphs. To do this a few preliminary considerations must be stated.

First, if the graph contains a chain then this is contracted to a single edge. The reduced graph has equivalent geometric characteristics. Each vertex, however, is now connected to at least three other vertices.

Second, it should be recognized that a proper criterion states only the minimum requirement. This is equivalent to the exclusion of all graphs containing nonplanar subgraphs. Therefore the graph must be connected by at least three vertices since Theorem 1.24 requires a nonplanar subgraph if it is not. From these considerations the nonplanar graphs (a) have only vertices where $\rho(a) \geqslant 3$, (b) contain no nonplanar subgraphs, and (c) are connected by at least three vertices.

It is assumed that with a given construction the nonplanar edge lies between vertices a and b. If nonplanar edge (a, b) is removed then in accordance with Menger's theorem there must remain a planar graph which contains a circuit through a and b. If there are several circuits then the circuit enclosing the most edges is selected. This circuit is a continuous closed curve which divides the graph into an inner and an outer component. Each component may be attached to the circuit at several vertices and may contain numerous transversals. Any outer transversal contains one and only one attachment to each of the arcs connecting vertices a and b. If it were attached in more than one place then a larger circuit could be selected. If it were attached on only one arc then the graph would be articulated.

Any inner transversal whose attachments are not separated by the vertex of an outer transversal can be transferred to the exterior. After the transfer it may be possible to select a larger circuit through vertices a and b. This process is continued until all irrelevant inner transversals are removed.

Parallel transversals add nothing to the ability of any component to make the graph nonplanar and are coalesced to a single transversal. Since the graph is finite all of the above processes must reach a conclusion. At this time the final maximal circuit and the resultant outer transversal have the relationship shown in part (a) of Figure 1.16.

There must be an inner transversal from arc (a, c, b) to arc (b, d, a) to make edge (a, b) nonplanar. There must be an inner transversal from arc (c, b, d) to arc (d, a, c) to assure that the inner component can not be transferred to the exterior. Parts (b) through (f) of Figure 1.16 show all of the unique configurations of the inner component which are possible with four or fewer attachments. (In part (c) vertex e can lie at vertex b and in part (d) vertex e can lie at vertex c but these special cases produce nothing new.)

Part (f) is the Kuratowski pentoid. It is left as an exercise to show that all other figures contain the hexoid.

Adding an attachment to any quadrant in any of the graphs in Figure 1.16 produces a graph which contains another of the graphs as a subgraph. Consequently there is no need to consider more than four attachments. This shows that all nonplanar graphs contain at least one of the Kuratowski graphs and this completes the proof of Theorem 1.25.

In 1736 Euler published a solution for a puzzle concerning a promenade over the bridges in a small town called Königsberg. His report, which is considered to be the origin of graph theory, also presented the following important formula for planar graphs.

Theorem 1.26 *The number of faces in a connected planar graph is given by* $f = e - v + 2$.

The theorem is conveniently proved by selecting an arbitrary maximal tree. For a single edge joining two vertices the formula gives the value $f = 1$ which corresponds to the infinite face. As the maximal tree is completed both

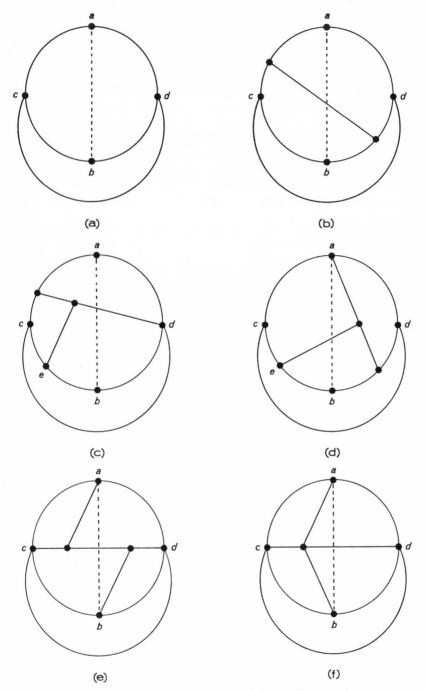

Figure 1.16 Possible configurations of the nonplanar graphs.

e and v increase by one each time an edge and a vertex are added so the value of f does not change. However, since the graph is planar whenever a chord is added it creates one and only one interior face. At the same time e increases by one but v does not change since the terminal vertex is present. So f increases by one and the formula gives the total number of faces. This includes the infinite face.

Corollary 1.26.1   *The circuit rank for a planar graph equals the number of interior faces.*

From the definition of the circuit rank it is true that

$$n(G) = e - v + 1 = f - 1 \qquad (1.18)$$

and this is the number of interior faces since the type of graphs under consideration have only one infinite face.

Theorem 1.27   *In any planar graph at least one vertex joins fewer than four edges or at least one face has fewer than four edges.*

Let v_ρ be the number of vertices with valence ρ and let f_σ be the number of faces with valence σ. (Face valence is the number of edges bounding a face.) Then the total quantities can be expressed as

$$v = \sum_\rho v_\rho$$
$$f = \sum_\sigma f_\sigma \qquad (1.19)$$

Since each edge connects two vertices and lies between two faces the number of edges is given by the relations

$$2e = \sum_\rho \rho v_\rho = \sum_\sigma \sigma f_\sigma \qquad (1.20)$$

Substituting each of these with Equation 1.19 into the Euler formula of Theorem 1.26 gives the equations

$$\sum_\rho v_\rho(2 - \rho) + 2\sum_\sigma f_\sigma = 4$$
$$\sum_\sigma f_\sigma(2 - \sigma) + 2\sum_\rho v_\rho = 4 \qquad (1.21)$$

and when these are added the result is

$$\sum_\rho v_\sigma(4 - \rho) + \sum_\sigma f_\sigma(4 - \sigma) = 8 \qquad (1.22)$$

For the left side of this equation to be positive it is necessary that for some $\rho < 4$ or some $\sigma < 4$ the corresponding v_ρ or f_σ is not zero. Hence there exists a vertex or a face with a valence less than four.

Since each edge appears at two vertices and on two faces the averages of the valences are given by

$$\rho_0 = \frac{2e}{v}$$

$$\sigma_0 = \frac{2e}{f}$$ (1.23)

and substitution in the Euler formula gives the equations

$$(\rho_0 - 2)v + 4 = 2f$$
$$(\sigma_0 - 2)f + 4 = 2v$$ (1.24)

Successively imposing the restrictions that $\rho_0 > 2$ and that $\sigma_0 > 2$ in the above equations immediately establishes the following two theorems.

Theorem 1.28 *If every vertex in a planar graph joins at least three edges then the numbers of faces and vertices are related by the formula $2f \geqslant v + 4$.*

Theorem 1.29 *If every face in a planar graph has at least three edges then the numbers of faces and vertices are related by the formula $2v \geqslant f + 4$.*

Equations 1.24 can be combined and written in the form

$$(\rho_0 - 2)(\sigma_0 - 2) = 4\left(\frac{f-2}{f}\right)\left(\frac{v-2}{v}\right)$$ (1.25)

and since f and v are positive nonzero numbers this relation can be expressed succinctly by the inequality

$$(\rho_0 - 2)(\sigma_0 - 2) < 4$$ (1.26)

This relationship between the averages of the valences is quite general and is true for any connected planar graph. If consideration is again restricted to graphs where $\rho_0 > 2$ or where $\sigma_0 > 2$ then the inequality immediately establishes the following two theorems.

Theorem 1.30 *If every vertex in a planar graph joins at least three edges then there is at least one face with fewer than six edges.*

Theorem 1.31 *If every face in a planar graph has at least three edges then there is at least one vertex which joins fewer than six edges.*

The following two lemmas describe some interesting properties of planar graphs and will be used in subsequent discussions.

Lemma 1.5 *A graph on a plane can be mapped to a sphere and a graph on a sphere can be mapped to a plane.*

Two opposite points on the sphere can be designated the north and south poles. The sphere is placed on the plane so that the south pole is tangent at an arbitrary point. A straight line drawn from the north pole to any point in the plane intersects the sphere in one and only one point. This establishes a one-to-one correspondence between points on the plane and points on the sphere. Such correspondence provides a unique mapping in either direction. This method is called *stereographic projection*. It should be observed that the infinite face in the planar graph becomes finite on the sphere.

Lemma 1.6   *A planar graph can be transformed isomorphically so that any desired face is the infinite face.*

The graph is mapped to a sphere which is then rotated to place the north pole within the desired face. By a stereographic projection this face becomes the infinite face.

If for a given planar graph there exists another graph such that each edge crosses one and only one edge of the given graph and each vertex lies in one face of the given graph then the second graph is called a *P-dual* of the first graph. Figure 1.17(a) shows the edge-to-edge correspondence and part (b) shows the vertex-to-face relationship.

The nomenclature signifies that the given graph is planar. An attempt to construct a *P*-dual of a Kuratowski graph indicates that the possession of a *P*-dual may be a unique characteristic of planar graphs. The proof of this fact, however, can not be based on a definition which requires the original graph to be planar.

In the five years following 1930 Whitney published a series of reports on graph theory in which he introduced and utilized the following ingenious algebraic definition of duality.

If every edge in a graph G_2 corresponds to an edge in a graph G_1 then

$$e(G_2) = e(G_1) \tag{1.27}$$

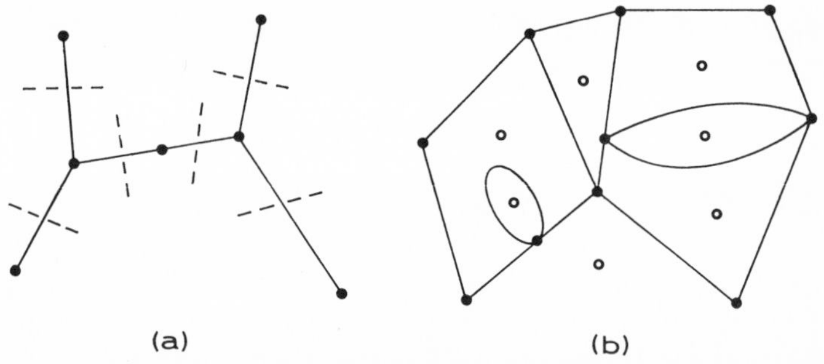

(a)   (b)

Figure 1.17   Edge and vertex relationships in a *P*-dual graph.

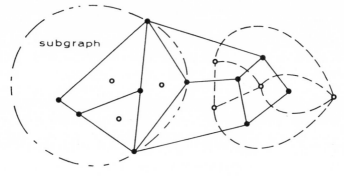

subgraph

complement

Figure 1.18 Definition of W-duality.

and for all corresponding subgraphs it is true that

$$e(g_2) = e(g_1) \qquad (1.28)$$

Also if all g_1 in G_1 satisfy the requirement

$$n(g_1) + r(G_2 - g_2) = r(G_2) \qquad (1.29)$$

then G_2 is called a *W-dual* of G_1. The origin of this puzzling definition is shown in Figure 1.18. The circuit rank $n(g_1)$ gives the number of interior faces in the subgraph. The cut-set rank $r(G_2 - g_2)$ equals one less than the number of vertices in a *P*-dual of the complement. The sum of these should equal one less than the number of vertices in G_2. This can be expressed as $r(G_2)$. So the Whitney definition of duality merely represents two ways of counting vertices. Figure 1.18 shows a planar graph but Equation 1.29 does not imply such a requirement. However, the above argument shows that every planar graph satisfies Equation 1.29 and establishes the following theorem.

Theorem 1.32 *Every P-dual is a W-dual.*

The converse of this theorem is not obviously true. It is conceivable that some nonplanar graphs may possess *W*-duals. It is proved later, however, that this does not happen. The following theorems state some interesting properties of dual graphs.

Theorem 1.33 *Every planar graph has a P-dual.*

This is conveniently proved by induction relative to the number of faces in the graph. The theorem is obviously true for one face since the dual graph has one interior and one exterior vertex joined by the required number of edges. It is necessary to show that if the construction is possible for a graph of f faces it can be done when the graph has $f + 1$ faces. In the most general case the additional face is created by adding a chain of edges to a circuit somewhere in the graph. To simplify the discussion the graph is transformed by two stereographic projections such that the face within the circuit becomes the infinite face. As shown in Figure 1.19(a) the P-dual graph has one interior vertex and one edge corresponding to each edge of the circuit. Part (b) shows how the dual graph is constructed when the additional face is attached. A vertex is added to the dual graph to correspond to the added face. Edges in the dual graph which correspond to edges lying between the points of attachment and which formerly joined the vertex of the dual graph corresponding to the infinite face now join the vertex in the added face. Edges with the required one-to-one correspondence are then added to the dual graph to join this vertex to the vertex in the infinite face. This completes the construction. The method applies equally well when there are pendant edges or loops present.

Theorem 1.34   *Every planar graph has a W-dual.*

It has already been shown that a P-dual to a planar graph meets the requirement for W-duality. The previous theorem shows that every planar graph has a P-dual. As a consequence, for every planar graph there exists another graph which is its W-dual. This rather trivial extension is needed in the logical development of a subsequent theorem.

Theorem 1.35   *P-duality is reciprocal.*

Reciprocity in this sense means that if G_2 is a P-dual of G_1 then G_1 is a P-dual of G_2. The edge-to-edge correspondence is obviously reciprocal so it must be shown only that each vertex of G_1 lies in a face of G_2. Figure 1.20 shows the three possible situations. In part (a) the vertex is interior to the graph. The faces cover the entire $360°$ and the edges of the dual graph form a circuit.

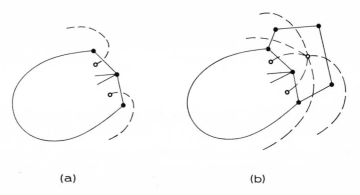

(a)   (b)

Figure 1.19   Adding a face to a P-dual graph.

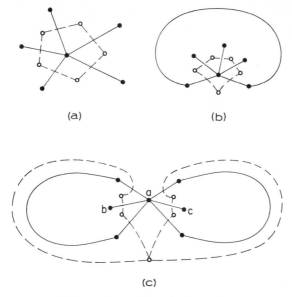

(a)

(b)

(c)

Figure 1.20   Reciprocity of *P*-duality.

The vertex of G_1 lies in a face of G_2. In part (b) the vertex is on an outer circuit. The edges of the dual graph which correspond to edges of G_1 which lie on the outer circuit are connected to the vertex of the dual graph which lies in the infinite face. Again a circuit is formed and the vertex of G_1 lies in a face of G_2. Figure 1.20(c) shows that if the vertex of G_1 is an articulation point then it is not enclosed by a circuit of G_2. However, it is the vertex in the infinite face of G_2 and the required vertex-to-face correspondence is maintained. It should be observed that to pass from vertex b to vertex c it is necessary to pass through vertex a so the *P*-dual of a separable graph is separable.

Theorem 1.36   *W-duality is reciprocal.*

In the Whitney criterion

$$n(g_1) + r(G_2 - g_2) = r(G_2) \qquad (1.30)$$

letting $g_1 = G_1$ and $g_2 = G_2$ produces the relation

$$n(G_1) = r(G_2) \qquad (1.31)$$

Subtracting each side of Equation 1.30 from

$$e(g_1) + e(G_2 - g_2) = e(G_1) \qquad (1.32)$$

and substituting Equation 1.31 gives

$$[e(g_1) - n(g_1)] + [e(G_2 - g_2) - r(G_2 - g_2)] = [e(G_1) - n(G_1)]$$
(1.33)

Using the fact that $e = r + n$ for any graph this reduces to

$$n(G_2 - g_2) + r(g_1) = r(G_1)$$
(1.34)

By selecting new subgraphs $g_1' = G_1 - g_1$ and $g_2' = G_2 - g_2$ this is converted to the form

$$n(g_2') + r(G_1 - g_1') = r(G_1)$$
(1.35)

Since g_1 is arbitrary this relation is reciprocal to the criterion of Equation 1.30.

If one lets $g_1' = G_1$ in the above equation it reduces to

$$n(G_2) = r(G_1)$$
(1.36)

Combining this with Equation 1.31 shows that for W-dual graphs the cut-set rank and circuit rank are interchanged.

Theorem 1.37 *A cut-set in a nonseparable graph corresponds to a circuit in both a P-dual and a W-dual graph.*

Parts (a) and (b) of Figure 1.20 and the associated discussion establish this theorem for P-duality. Further, part (c) of the same figure shows why it is necessary for the graph to be nonseparable. It is left as an exercise to extend the proof from a basic cut-set to a general cut-set.

The proof for W-duality must necessarily use the Whitney criterion. Let S_1 be the set of edges in G_1 which form a cut-set. Let S_2 be the corresponding set of edges in the W-dual graph G_2. Recalling that for any graph $r = v - p$ where v is the number of vertices and p is the number of parts permits Equation 1.35 to be written as

$$n(S_2) + [v(G_1) - 2] = [v(G_1) - 1]$$
(1.37)

This quickly reduces to $n(S_2) = 1$ which shows that S_2 has one circuit. If any subgraph s_1 of the cut-set S_1 is deleted then the graph G_1 remains in one part and Equation 1.35 becomes

$$n(s_2) + [v(G_1) - 1] = [v(G_1) - 1]$$
(1.38)

This immediately reduces to $n(s_2) = 0$. Hence S_2 has a circuit and no subgraph has a circuit so S_2 is a circuit.

Theorem 1.38 *A circuit in a nonseparable graph corresponds to a cut-set in both a P-dual and a W-dual graph.*

This is a direct consequence of the last three theorems.

Theorem 1.39   *If a graph has a W-dual it is planar.*

If graphs conformal to the Kuratowski graphs have no *W*-duals then the only graphs which have *W*-duals are planar graphs. Therefore to prove this theorem it is sufficient to prove that the Kuratowski graphs have no *W*-duals. This is very easy. The hexoid contains 9 undivided quadrilaterals. According to Theorem 1.38 the dual graph must contain 9 vertices with a valence of 4. This requires 18 edges. However, the edge-to-edge correspondence provides only 9 edges in the dual graph so the construction is impossible. The pentoid contains 10 triangles so its dual graph must contain 10 vertices with a valence of 3. This requires 15 edges. The edge-to-edge correspondence provides only 10 edges so the construction again is impossible. As a matter of logic it should be observed that this proof tacitly assumes the rather obvious lemma that a graph has no *W*-dual if any of its subgraphs fail to possess a *W*-dual. Also, the extension from a Kuratowski graph to a general nonplanar graph involves the substitution of chains of edges for single edges plus the addition of planar and nonplanar subgraphs. It is left as an exercise to show that these extensions do not invalidate the theorem.

It has been show that every *P*-dual graph meets the requirement for *W*-duality and that if a graph has a *W*-dual it is planar. But all planar graphs have a *P*-dual. The conclusion is that *W*-duality contains *P*-duality. It is no longer necessary to specify the type of duality. Two of the previous theorems can be modified to state that (a) if a graph is planar it has a dual graph and (b) if a graph has a dual graph it is planar. These are the necessary and sufficient conditions for the following important theorem.

Theorem 1.40   *A graph is planar if and only if it has a dual graph.*

By means of isomorphic transformations a given planar graph can be represented on the plane in numerous ways. This introduces the question as to whether or not all of the corresponding dual graphs are isomorphic. A few immediate conclusions can be drawn. First, transformations using stereographic projection produce graphs having isomorphic dual graphs since the projection maintains the same vertex-to-face correspondence. Second, dual graphs are not unique if the graph is separable. This is illustrated in Figure 1.21 where the separable graph consists of only two loops. An isomorphic change in the original graph produces an interchange of connections from vertex *a* to vertex *b* so the dual graphs are not isomorphic. It is necessary therefore to investigate the various possible isomorphic changes in planar graphs and to determine the effect which each has on the configuration of the dual graph. At this point a suggested exercise is to show that the dual graph of a nonseparable graph is nonseparable.

For graphs connected at two vertices there are two possible isomorphic changes. These are shown in Figure 1.22. In a *translation* the subgraph *g* is

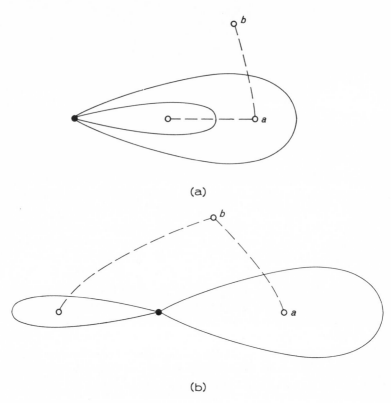

(a)

(b)

Figure 1.21   Simple graphs with dual graphs which are not isomorphic.

connected on the alternate side of the subgraph $G - g$ but the internal connections remain the same. In an *inversion* of g its position relative to subgraph $G - g$ is maintained but the mirror image of the internal connections is used.

It is easily deduced that regardless of which graph is translated the result is the same and that an even number of translations produces the original graph. Therefore the most general case is a single translation.

Similarly it is easily concluded that either graph may be inverted since the results are equivalent through a stereographic projection, and that an even number of inversions produces either the original graph or an isomorphism which is equivalent through a stereographic projection. Hence it is sufficient to consider a single inversion.

Continuing the investigation into combinations of translation and inversion it is found that the only operations which produce truly unique isomorphisms are (a) a translation, (b) an inversion, and (c) a translation plus an inversion. Figure 1.23 shows these operations and the corresponding

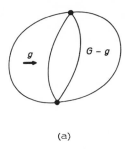

(a)

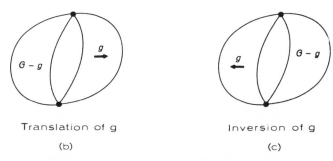

Translation of g   Inversion of g

(b)   (c)

Figure 1.22 Translation and inversion of a subgraph.

dual graphs. In part (b) the inversion has the effect of interchanging the connections to vertices a and b of the edges of the dual graph emanating from subgraph g. The dual graph is not isomorphic to the dual graph of the original graph. It is said to be 2-*isomorphic* to the original dual graph. It is an interesting exercise to show that there is a circuit-to-circuit correspondence in 2-isomorphic graphs.

In part (c) of Figure 1.23 the connections of the dual edges to vertices a and b remain the same so a translation produces an isomorphic dual graph.

In part (d) the connections from g to vertices a and b are again interchanged so the dual graph is 2-isomorphic to the original dual graph.

An attempt to translate or invert a graph connected at more than two vertices is impossible since it violates the principle of Figure 1.13. The preceding discussion establishes the following theorem.

Theorem 1.41 *A 2-isomorphic dual graph is produced if and only if there is an odd number of inversions of the original graph.*

In addition to the theorems of Kuratowski and Whitney there is a third method of characterizing planar graphs. It is due to MacLane and it is concerned with the structure of the circuits in the graph.

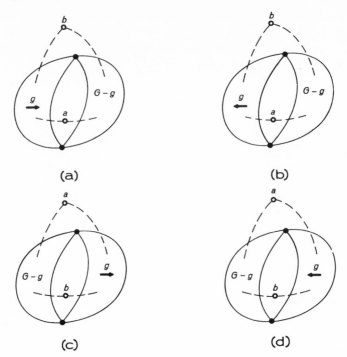

Figure 1.23   Creation of a 2-isomorphic dual graph.

Theorem 1.42   *A graph is planar if and only if it has a set of basis circuits such that every edge lies on two basis circuits and every circuit is the sum mod 2 of basis circuits.*

The basis circuits of this theorem are not necessarily the same as basic circuits. They can be envisioned reservedly as those circuits which bound faces in the graph. The sum mod 2 means only that an edge is deleted if it appears on two basis circuits. To prove the necessity of the conditions it is assumed that the graph is planar and hence has a dual graph. In a given representation each basis circuit defines a face in the graph. Since the graph has a dual graph there is an edge-to-edge correspondence and each vertex of the dual graph lies in a face of the original graph. Hence each edge of the original graph lies on two basis circuits.

If a given circuit in the graph has no inner transversal then it is a basis circuit. If it has one inner transversal then it contains two basis circuits and the edges of the transversal lie on both basis circuits but do not appear in the given circuit. If the given circuit contains numerous inner transversals then it can be constructed by adding one basis circuit at a time and the above rule is followed at each step.

The sufficiency of the conditions can conveniently be established by relating them to the requirements of W-duality. Take subgraph g_1 as any set of basis circuits which form a circuit in G_1. If this set contains c_1 basis circuits then

$$r(G_2 - g_2) = v(G_2 - g_2) - 1$$
$$= [v(G_2) - c_1] - 1 \qquad (1.39)$$
$$r(G_2) = v(G_2) - 1$$

Since g_1 contains c_1 circuits $n(g_1) = c_1$ and for any g_1 it is true that

$$n(g_1) + r(G_2 - g_2) = r(G_2) \qquad (1.40)$$

Hence a graph constructed in accordance with the MacLane conditions meets the requirements for W-duality and by Theorem 1.39 it must be planar.

Although the criteria of Kuratowski, Whitney, and MacLane are easily envisioned their application to a complicated graph is quite difficult. To alleviate this situation numerous routines have been developed. These usually incorporate the structure of the graph as a matrix or as an ordering of the vertices along selected paths. A systematic procedure is then followed to scan the graph and to determine whether or not it meets certain requirements. Although these methods have significant practical value they are quite esoteric and can not be included here. A report by Lempel, Even, and Cederbaum which is listed as a reference outlines a recently developed procedure and gives a bibliography of some previous ones.

This chapter opened with an allusion to the impetus given graph theory by persons working on the four-color conjecture. It is perhaps interesting to conclude with a brief summary of some of their results. Since map coloring is only remotely related to network theory these are stated without proof. For a rather complete discussion reference is made to the recent book by Ore. All statements apply to the face coloration of planar connected graphs.

A graph can be colored in only two colors when all of its vertex valences are even. This happens (a) if it is formed by the intersection of a finite set of closed curves or (b) if it is constructed from a finite set of infinite straight lines.

A graph can be colored in three colors (a) if all of its faces are triangles or (b) if all of its face valences are even and all of its vertex valences equal 3.

A graph can be colored in four colors (a) if all of its face valences are less than 5, (b) if it contains a Hamilton circuit, or (c) if it can not be separated by less than 4 vertices.

REFERENCES

1. O. Ore, *Theory of Graphs*, American Mathematical Society, Providence, 1962.
2. C. Berge, *The Theory of Graphs*, Wiley, New York, 1962.

3. R. G. Busacker and T. L. Saaty, *Finite Graphs and Networks*, McGraw-Hill, New York, 1965.
4. L. R. Ford and D. R. Fulkerson, *Flows in Networks*, Princeton University Press, Princeton, 1962.
5. J. Riordan, *An Introduction to Combinatorial Analysis*, Wiley, New York, 1958.
6. O. Ore, *The Four-Color Problem*, Academic Press, New York, 1967.
7. H. Whitney, The Abstract Properties of Linear Dependence, *American Journal of Mathematics*, Volume 57, 1935, pp. 509–533.
8. H. Whitney, Congruent Graphs and the Connectivity of Graphs, *American Journal of Mathematics*, Volume 54, 1932, pp. 150–168.
9. H. Whitney, Non-Separable and Planar Graphs, *Transactions of the American Mathematical Society*, Volume 34, 1932, pp. 339–362.
10. C. Kuratowski, Sur le problème des courbes gauches en Topologie, *Fundamenta Mathematicae*, Volume 15, 1930, pp. 271–283.
11. H. Whitney, Planar Graphs, *Fundamenta Mathematicae*, Volume 21, 1933, pp. 73–84.
12. H. Whitney, 2-Isomorphic Graphs, *American Journal of Mathematics*, Volume 55, 1933, pp. 245–254.
13. S. MacLane, A Combinatorial Condition for Planar Graphs, *Fundamenta Mathematicae*, Volume 28, 1937, pp. 22–32.
14. A. Lempel, S. Even, and I. Cederbaum, *An Algorithm for Planarity Testing of Graphs*, *Theory of Graphs*, Gordon and Breach, New York, 1967.

2
NETWORK FUNCTIONS

One of the more interesting and fruitful scientific endeavors of mankind has been the search for the conditions under which an extreme situation occurs. The ancient Greek geometers conjectured that a circle and a sphere enclose the greatest area and volume respectively for a line of fixed length and a surface of fixed area. Thus at an early date they considered what are known as *isoperimetric problems*.

Interest in this type of problem was revived and stimulated by the discovery of Fermat in the seventeenth century that when a ray of light travels from one fixed point to another it seeks the path which minimizes the time of travel. This principle is very general and from this fact it is possible to deduce the fundamental laws of geometric optics such as those governing reflection and refraction.

The fact that in this case nature appears to do things with the greatest economy of effort led some scientists and philosophers to attach religious significance to this principle and to extend their investigations into other areas. In the eighteenth century Euler developed some fundamental concepts of a branch of mathematics known as the *calculus of variations*. In this discipline requirements are placed on a function so as to provide an extreme value for an integral involving the function. In 1788 Lagrange adapted the calculus of variations to problems in the dynamics of bodies in motion and produced some fundamental results. A climax to this work was reached in 1834 when Hamilton announced the principle which now bears his name and which occupies a unique place in all of science for its elegance and simplicity.

All of these achievements were related only to mechanical systems since the science of electricity was embryonic at the time. During the subsequent advance of electrical technology the corresponding mathematical analysis relied heavily on Kirchhoff's laws rather than the results of classical dynamics. There are two good reasons. First, electrical systems are frequently more complex topologically and an analysis based on graph theory and the properties of networks is undeniably appropriate. Second, in classical dynamics the mass is constant and the force of a dissipative effect is usually assumed to be proportional to the velocity. Therefore classical analysis deals primarily with linear effects. Due to magnetic saturation it was recognized at an early date that electrical inductors are frequently nonlinear. Modern technology has produced nonlinear resistors and capacitors. It would appear that the beautifully succinct principles of classical dynamics do not apply to common electrical networks.

Fortunately this is not true. In 1951 Millar and Cherry showed how the concepts of power and energy can be revised so that the principles of classical dynamics apply equally well to nonlinear networks. Part of their work contemplated networks of similar elements but with only one terminal pair. This was then generalized to networks of similar elements which have multiple terminal pairs. This has been extended further by the introduction of a hybrid network function which has much greater flexibility than those employed in the classical theory.

The primary purpose of this chapter is to present methods for generating a set of equations whose solutions give an answer to the network problem The selection of appropriate variables is discussed along with the properties of numerous mathematical functions of these variables. The equations must necessarily incorporate the characteristics of the devices in the network and they must also reflect the manner in which the devices are connected. It is to be expected, therefore, that portions of graph theory as discussed in the first chapter will play a prominent part in the development.

This chapter opens with a brief summary of classical dynamics. This is then extended to include electrical elements and nonlinear devices. It is shown that nonlinear electrical networks conform to the principles of classical dynamics and display extreme values for some of their network functions. Kirchhoff's laws are then combined with some fundamental results of graph theory to predict the number of variables and equations which one may expect. Following this is the central discussion on generation of equations. Some new methods are presented. It is shown by example and comparison that the classical methods of Lagrange and Hamilton yield equations similar to those derived by Kirchhoff's laws and the newer methods. Since the choice of method should be based on mathematical convenience, a method yielding equations in the simplest form is developed.

CLASSICAL DYNAMICS

Euler was probably the first to contribute in a unique and significant manner to the general problem of finding the conditions for an extremum. He considered an integral of the form

$$I = \int_a^b f(x, y, y') \, dx \tag{2.1}$$

where $y = y(x)$ and y' is the derivative with respect to x. The problem is to determine the form of $y(x)$ which gives an extreme value for I. To do this Euler created a variation in $y(x)$ and replaced it with $y(x) + \alpha\eta(x)$ where $\eta(x)$ is an arbitrary, fixed, finite function with the property that $\eta(a) = \eta(b) = 0$. If f is well behaved then the definite integral is now a function of α and the values of the variables at the end points of the interval. Assuming that the end points are fixed then the variation in the value of the integral around the point where $\alpha = 0$ is given by

$$\frac{\partial I}{\partial \alpha}\bigg|_{\alpha=0} = \int_a^b \left[\frac{\partial f}{\partial y}\eta + \frac{\partial f}{\partial y'}\eta'\right] dx$$

$$= \left[\frac{\partial f}{\partial y'}\eta\right]_a^b - \int_a^b \left[\frac{\partial f}{\partial y} - \frac{d}{dx}\frac{\partial f}{\partial y'}\right]\eta\, dx \qquad (2.2)$$

where the second equation is derived through an integration by parts. If the integral I is to have an extreme value then $y(x)$ must be selected so as to make this variation equal to zero. Vanishing of the first term is assured by the properties of $\eta(x)$. However, since $\eta(x)$ is arbitrary except at the end points the second term is zero if and only if

$$\frac{d}{dx}\frac{\partial f}{\partial y'} - \frac{\partial f}{\partial y} = 0 \qquad (2.3)$$

This famous result is called the *Euler equation*. A function $y(x)$ which satisfies this equation gives what is known as a *stationary* value to the integral I. This means that the value of the integral is constant for small variations in $y(x)$. This does not always create an extreme condition. If the derivative of I with respect to α should have a double root then there is a saddle point. This appears as a plateau with a rise on one side and a drop on the other. The Euler equation therefore must be regarded as only a necessary condition. The sufficient condition requires an investigation of the behavior of the first derivative on both sides of the null point or, equivalently, an evaluation of the second derivative.

Lagrange derived a relation similar to the Euler equation from entirely different considerations. He was concerned with problems in mechanics and his analysis is based on concepts of energy and force. As an example, the inertial force is given by Newton's law,

$$F_m = m\ddot{x} = m\frac{d\dot{x}}{dt} \qquad (2.4)$$

The energy acquired by a body when it is accelerated from rest is called the *kinetic energy* and is given by

$$T = \int_0^x F_m\, dx = \int_0^x m\frac{d\dot{x}}{dt}\, dx = \int_0^{\dot{x}} m\dot{x}\, d\dot{x} = \tfrac{1}{2}m\dot{x}^2 \qquad (2.5)$$

The inertial force can be derived from the kinetic energy by the relation

$$F_m = \frac{d}{dt}\frac{dT}{d\dot{x}} \tag{2.6}$$

If there is a force such as elastic restraint or gravitation which is single valued and dependent only on position so that

$$F_k = F_k(x) \tag{2.7}$$

then it is possible to define uniquely a *potential energy* which is given by

$$U = \int_0^x F_k(x)\,dx \tag{2.8}$$

and the force which creates this energy can be found quite easily from the relation

$$F_k = \frac{dU}{dx} \tag{2.9}$$

If there is a damping force it is usually assumed to depend linearly on the velocity and therefore is expressed by

$$F_d = a\dot{x} \tag{2.10}$$

When the body moves at a velocity $\dot{x}$ the power which is dissipated is given by

$$P_d = F_d\dot{x} = a\dot{x}^2 \tag{2.11}$$

Defining the *dissipation function* G as one half of this power so that

$$G = \tfrac{1}{2}a\dot{x}^2 \tag{2.12}$$

permits the damping force to be derived quite easily by the relation

$$F_d = \frac{dG}{d\dot{x}} \tag{2.13}$$

In the general case where there is an external force F applied to the body it is true that

$$F_m + F_k + F_d = F \tag{2.14}$$

A substitution of Equations 2.6, 2.9, and 2.13 produces the alternate form

$$\frac{d}{dt}\frac{dT}{d\dot{x}} + \frac{dU}{dx} + \frac{dG}{d\dot{x}} = F \tag{2.15}$$

This equation slightly resembles the Euler equation. It can be placed into a very similar form by introducing a function known as the *Lagrangian* which is defined by

$$L(x, \dot{x}) = T(\dot{x}) - U(x) \tag{2.16}$$

This function is the excess of the kinetic energy over the potential energy and therefore it has no simple physical interpretation. However, it permits Equation 2.15 to be written in the form

$$\frac{d}{dt}\frac{\partial L}{\partial \dot{x}} - \frac{\partial L}{\partial x} = F - \frac{\partial G}{\partial \dot{x}} \tag{2.17}$$

This important result is known as the *Lagrange equation*. It represents a summation of forces. Apart from its theoretical importance it provides a means for determining the differential equation describing a system when only the Lagrangian and the dissipation function are known.

The similarity between the left side of the Lagrange equation and Equation 2.3 suggests that the integral of the Lagrangian has stationary properties. This is true in a special case. If the applied force equals the damping force or if both are zero then Equation 2.17 reduces to

$$\frac{d}{dt}\frac{\partial L}{\partial \dot{x}} - \frac{\partial L}{\partial x} = 0 \tag{2.18}$$

This is the Euler equation and any $x(t)$ which satisfies this equation must provide a stationary value for the integral of $L(x, \dot{x})$. This fact is usually stated in the simple form

$$\delta \int_0^t L \, dt = 0 \tag{2.19}$$

which is known as *Hamilton's principle*. It states that the time integral of the Lagrangian is stationary for the particular motion $x(t)$ which is in agreement with the Lagrange equation. But this equation is only a statement of the summation of forces in accordance with Newton's law and the existence of a potential energy field. A motion $x(t)$ which satisfies these laws is precisely the one which is found in the world about us. The conclusion from a philosophical view is that the existing physical laws provide a unique framework within which nature can do things in an optimum manner. For example, should the inertial force depend on the velocity or the third derivative then it is impossible to make the Lagrange equation look like the Euler equation and the integral of the Lagrangian would not be stationary.

The conditions which provide Equation 2.18 define what is known as a *conservative system*. In such a system the *Hamiltonian* function

$$H(x, \dot{x}) = T(\dot{x}) + U(x) \tag{2.20}$$

is constant since it represents the total energy. A direct substitution into Equation 2.19 gives the alternate representations

$$\delta \int_0^t 2T \, dt = -\delta \int_0^t 2U \, dt = 0 \tag{2.21}$$

The time integral of $2T$ is called the *action*. Further investigation shows that the stationary value of the action is normally a minimum. This is known as the *principle of least action* and was formulated by Maupertuis in the eighteenth century.

The previous derivation is the simplest possible since it considers a single particle moving in only the x direction. A simple extension is to introduce the time t as an explicit variable. A comparison of the Euler and Lagrange equations shows immediately that t can appear explicitly in the Lagrangian without disturbing the results. Since the kinetic energy T is derived from Newton's law and since the mass is considered to be constant then it would be an unusual situation indeed if T contains t explicitly. However, it is quite conceivable that the force field which produces the potential energy may vary in time so it is realistic to assume that U may contain t explicitly. Systems whose parameters vary in time are said to be *rheonomic*. Otherwise they are called *scleronomic* systems.

Another simple generalization is to extend the scope of the problem to include motion in three dimensions. If the x, y, z axes are mutually perpendicular then all forces and inertial effects along these axes are independent. The Lagrangian function can now contain as many as seven explicit variables. In addition to the time t there are the three coordinates plus the three velocities. There are three Lagrange equations representing summations of forces in the three directions. The result is three differential equations which can, in theory, be solved to give the position as a function of the time.

If there are several independent particles in the system then there are three Lagrange equations for each mass. For a system of n particles there are required $3n$ coordinates to describe their positions and there are as many Lagrange equations.

In many practical problems such as those involving the rotation or revolution of particles the mathematical expressions are greatly simplified when specially appropriate coordinate systems are used. Common examples are the spherical and cylindrical coordinate systems. It is desirable to determine the form of the Lagrange equations under a general coordinate transformation. The result is somewhat astonishing. For particle j the transformation has the general form

$$x_j = x_j(t, q_1, \ldots, q_{3n})$$
$$y_j = y_j(t, q_1, \ldots, q_{3n}) \tag{2.22}$$
$$z_j = z_j(t, q_1, \ldots, q_{3n})$$

and it is shown in many books on advanced dynamics that under such a transformation Equation 2.17 becomes

$$\frac{d}{dt}\frac{\partial L}{\partial \dot{q}_s} - \frac{\partial L}{\partial q_s} = F_s - \frac{\partial G}{\partial \dot{q}_s} \tag{2.23}$$

This is the Lagrange equation in generalized coordinates. But the form has not changed; the Lagrange equation is invariant under a general coordinate transformation. This is one of its primary advantages. In a Cartesian coordinate system the Lagrange equation represents a summation of forces. In a generalized coordinate system this need not be true, yet the particular form of the equation is preserved although its physical significance may change.

Employing an analogy with the situation in a Cartesian coordinate system Hamilton introduced a *generalized momentum* whose component corresponding to q_s is defined by

$$p_s = \frac{\partial L}{\partial \dot{q}_s} \qquad (2.24)$$

This can be written as

$$\dot{p}_s = \frac{d}{dt} \frac{\partial L}{\partial \dot{q}_s} \qquad (2.25)$$

If the system is conservative and not subject to external forces then a substitution of the appropriate Lagrange equation gives

$$\dot{p}_s = \frac{\partial L}{\partial q_s} \qquad (2.26)$$

Using Equations 2.24 and 2.26 the total differential of L can be written as

$$dL = \sum_s (\dot{p}_s \, dq_s + p_s \, d\dot{q}_s) + \frac{\partial L}{\partial t} \, dt \qquad (2.27)$$

In generalized coordinates the Hamiltonian function is defined as

$$H = \sum_s p_s \dot{q}_s - L \qquad (2.28)$$

By the use of Equation 2.27 the total differential of H can be written as

$$dH = \sum_s (\dot{q}_s \, dp_s - \dot{p}_s \, dq_s) - \frac{\partial L}{\partial t} \, dt \qquad (2.29)$$

From this it is apparent that

$$\frac{\partial H}{\partial t} = -\frac{\partial L}{\partial t} \qquad (2.30)$$

and in addition

$$\dot{q}_s = \frac{\partial H}{\partial p_s}$$

$$\qquad (2.31)$$

$$\dot{p}_s = -\frac{\partial H}{\partial q_s}$$

These are Hamilton's *canonical equations*. They are valid only in a conservative system but both the Lagrangian and Hamiltonian functions may contain the time t explicitly. In a system of n unrestrained particles the Lagrange equations normally give $3n$ second-order differential equations. In a similar situation the Hamilton equations yield $6n$ first-order differential equations. This is perhaps their primary advantage. It is interesting to observe that the small world of atomic processes and the large world of celestial activities are both conservative and the Hamiltonian formulation is frequently used. Conversely, systems encountered in typical engineering applications are usually dissipative and almost always forced. They require the Lagrange equations in their more general form. Several interesting properties of the general Hamiltonian function are given in the following theorems.

Theorem 2.1   *In a conservative, scleronomic system the generalized Hamiltonian is constant.*

The total time derivative of H is expressed as

$$\frac{dH}{dt} = \sum_s \left(\frac{\partial H}{\partial q_s} \dot{q}_s + \frac{\partial H}{\partial p_s} \dot{p}_s \right) + \frac{\partial H}{\partial t} \tag{2.32}$$

but a substitution of the canonical equations reduces this to

$$\frac{dH}{dt} = \frac{\partial H}{\partial t} \tag{2.33}$$

Hence if the time t does not appear explicitly in H then the Hamiltonian is a constant of the motion.

Theorem 2.2   *If the potential energy is independent of the generalized velocities in a system described by a fixed coordinate system then the generalized Hamiltonian is the total energy.*

The kinetic energy of a system expressed in a coordinate system which is independent of the time has the form

$$T = \sum_k \sum_s \tfrac{1}{2} A_{ks} \dot{q}_k \dot{q}_s \tag{2.34}$$

and the derivative with respect to a generalized velocity is

$$\frac{\partial T}{\partial \dot{q}_s} = \sum_k A_{ks} \dot{q}_k \tag{2.35}$$

This leads to the relation

$$\sum_s \frac{\partial T}{\partial \dot{q}_s} \dot{q}_s = \sum_s \sum_k A_{ks} \dot{q}_k \dot{q}_s = 2T \tag{2.36}$$

which is due to Euler. Since the potential energy U does not contain a velocity Equation 2.24 can be written as

$$\frac{\partial T}{\partial \dot{q}_s} = p_s \qquad (2.37)$$

and substituting this into Equation 2.36 leads to the result

$$H = \sum_s p_s \dot{q}_s - L$$

$$= 2T - L = T + U \qquad (2.38)$$

so under the stipulated conditions the Hamiltonian function is the total energy of the system. It is not required that the time t be absent from the potential energy so the theorem applies also to rheonomic systems.

The solution of the Lagrange equations for a conservative system is simplified whenever a coordinate q_s does not appear explicitly in the Lagrangian. In this case the equation reduces to

$$\frac{d}{dt}\frac{\partial L}{\partial \dot{q}_s} = 0 \qquad (2.39)$$

which can be integrated immediately to give

$$p_s = \frac{\partial L}{\partial \dot{q}_s} = \text{constant} \qquad (2.40)$$

This equation can be solved for $\dot{q}_s$ and the result can be substituted in L to eliminate q_s and $\dot{q}_s$ entirely from the problem. Whenever this happens the variable q_s is called a *cyclic* or *ignorable* coordinate.

There are many books on mechanics which give examples to illustrate the various methods which are presented here so this discussion is closed with a single example. Figure 2.1 shows a pendulum whose length can be varied.

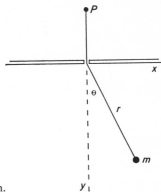

Figure 2.1   A pendulum of variable length.

Since there is one mass in motion three coordinates are needed to express its motion. However, there are two constraints. There is no force in the z direction so the particle must move in only the x, y plane. This is equivalent to a constraint and can be expressed by the relation

$$z = \text{constant} \tag{2.41}$$

Although it is assumed that the length of the pendulum may vary in time it is always true that

$$x^2 + y^2 = r^2 \tag{2.42}$$

As a consequence of the above constraints only one generalized coordinate is needed and this can conveniently be the angle θ. There is also only one Lagrange equation.

Since $x = r \sin \theta$ and $y = r \cos \theta$ the kinetic energy is given by

$$
\begin{aligned}
T &= \tfrac{1}{2} m (\dot{x}^2 + \dot{y}^2) \\
&= \tfrac{1}{2} m [(\dot{r} \sin \theta + r \cos \theta \dot{\theta})^2 + (\dot{r} \cos \theta - r \sin \theta \dot{\theta})^2] \\
&= \tfrac{1}{2} m (\dot{r}^2 + r^2 \dot{\theta}^2)
\end{aligned}
\tag{2.43}
$$

and the potential energy is expressed as

$$U = -mgr \cos \theta \tag{2.44}$$

so the Lagrangian is

$$L = T - U = \tfrac{1}{2} m (\dot{r}^2 + r^2 \dot{\theta}^2) + mgr \cos \theta \tag{2.45}$$

and the Lagrange equation is

$$\frac{d}{dt} \frac{\partial L}{\partial \dot{\theta}} - \frac{\partial L}{\partial \theta} = \frac{d}{dt} (mr^2 \dot{\theta}) + mgr \sin \theta = 0 \tag{2.46}$$

This is a nonlinear equation in the dependent variable θ. It appears in a more tractable form when it is assumed that the vibrations are so small that $\sin \theta$ can be replaced with θ. In this case the equation reduces to

$$\ddot{\theta} + \frac{2\dot{r}}{r} \dot{\theta} + \frac{g}{r} \theta = 0 \tag{2.47}$$

This is a second order equation typical of those which arise from a Lagrange equation. If r is constant then the damping is zero and the solution is simple harmonic motion. However, since r is always positive if P moves downward then $\dot{r} > 0$ and there is positive damping. The amplitude of the vibration decreases. Conversely, if P moves upward the amplitude increases. This may be verified very easily with quite primitive equipment.

VARIOUS TYPES OF LAGRANGIANS

The preceding development of the Lagrange equation was based on the generation of the kinetic energy of motion by forces derived from various sources. In particular the force derived from a potential field was assumed to depend only on position. This is a restricted situation and this section presents some generalizations. However, in all of this discussion the Lagrangian will remain nondissipative and the extensions will be limited to the treatment of different potential energy sources and the discovery of some new devices.

The discussion is simplified by introducing a *Lagrangian operator* which is defined as

$$\mathscr{L}_s = \left(\frac{d}{dt} \frac{\partial}{\partial \dot{x}_s} - \frac{\partial}{\partial x_s} \right) \tag{2.48}$$

and since kinetic energy is not a function of position Newton's law can be written in the simpler form

$$\mathscr{L}_s T = y_s \tag{2.49}$$

where y_s is the generalized force along the generalized coordinate x_s. (The notation is changed because electrical systems are considered later and the electrical charge is traditionally designated by q.) If there exists a function $U(t, x_1, \ldots, \dot{x}_1, \ldots)$ such that each generalized force internal to the system can be calculated from the relation

$$\mathscr{L}_s U = y_s \tag{2.50}$$

then it is certainly true that

$$\mathscr{L}_s L = 0 \tag{2.51}$$

if L is again defined by $L = T - U$. Thus the Lagrange equation remains valid when the Lagrangian is extended to include force fields which depend on velocities. It is significant that, with the notable exception of Coulomb friction, all of the macroscopic forces of classical physics can be derived from such a generalized potential function. Coulomb friction produces heat which is not expressible in macroscopic coordinates or velocities. It can be represented in an approximate fashion as a component of the dissipation function.

Theorem 2.2 states that the Hamiltonian equals the total energy when the potential function is independent of the velocities. The following theorem gives an evaluation of the Hamiltonian in a frequently occurring case wherein the potential function is dependent on the velocities.

Theorem 2.3 *If the velocity-dependent potential function in a Lagrangian varies linearly with the velocities then the Hamiltonian gives the sum of the kinetic energy and that portion of the potential energy derived only from position.*

Under the assumed condition the Lagrangian can be written as

$$L = T - U_0 - \sum_s a_s \dot{x}_s \qquad (2.52)$$

where U_0 is a function only of position and a_s may contain coordinates. The Hamiltonian may be expressed by the various relations

$$H = \sum_s \dot{x}_s \frac{\partial L}{\partial \dot{x}_s} - L$$

$$= \sum_s \dot{x}_s \left[\frac{\partial T}{\partial \dot{x}_s} - \frac{\partial U_0}{\partial \dot{x}_s} - a_s \right] - \left[T - U_0 - \sum_s a_s \dot{x}_s \right]$$

$$= \sum_s \dot{x}_p \frac{\partial T}{\partial \dot{x}_s} - [T - U_0] \qquad (2.53)$$

The substitution of Equation 2.36 reduces this to

$$H = T + U_0 \qquad (2.54)$$

and this establishes the theorem.

Regardless of the form of the Lagrangian if Equation 2.51 is satisfied then the system is conservative since there is no provision for external forces or the dissipation of energy. There are two general types of devices which are conservative. In one type the energy is stored within the device and is released when the environment requests it. This type of device is called a *reactive* or *energic* element. The other type of nondissipative device is transmissive. It neither stores nor dissipates energy but acts as a medium for transferring energy from one location to another. This type is called a *nonenergic* element. In the class of mechanical devices a typical nonenergic element is the simple lever.

The simplest type of energic action is the uniform gravitational field. For a vertical coordinate the potential function is given by $U = mgx$ and the Lagrangian is $L = \frac{1}{2}m\dot{x}^2 - mgx$. The requirement that $\mathscr{L}L = 0$ provides the equation $\ddot{x} = -g$. Another simple, energic, mechanical element is the linear spring or elastic effect where the force is directly proportional to a displacement. The potential function is $U = \frac{1}{2}kx^2$ and the Lagrangian is $L = \frac{1}{2}m\dot{x}^2 - \frac{1}{2}kx^2$. The equation $\mathscr{L}L = 0$ yields $m\ddot{x} = -kx$. These results are the usual differential equations describing these simple systems.

The inductor and capacitor are energic electrical devices. The linear inductor has properties defined by

$$v = l\ddot{q}$$

$$i = \frac{\phi}{l} \qquad (2.55)$$

and these can be derived from the respective Lagrangians

$$T_l = \frac{l\dot{q}^2}{2}$$

$$U_l = \frac{\phi^2}{2l}$$

(2.56)

The generalized coordinates in electrical systems are the charge q and the total magnetic flux ϕ. Such usage eliminates integrals from the equations and permits the expressions to assume the form of the Euler equation. Equation 2.55 shows that the corresponding generalized forces are the voltage v and the current i respectively. It should be observed that the inductor has inertial properties when described by the flow of charge but it appears to be elastic when the magnetic flux is used as the coordinate. The linear capacitor is defined by

$$i = c\ddot{\phi}$$

$$v = \frac{q}{c}$$

(2.57)

where the flux is strictly a mathematical concept and its Lagrangian forms are

$$T_c = \frac{c\dot{\phi}^2}{2}$$

$$U_c = \frac{q^2}{2c}$$

(2.58)

Another energic electrical network element is a set of coupled coils. Their mutual effect on each other is described by

$$v_1 = m\frac{di_2}{dt}$$

$$v_2 = m\frac{di_1}{dt}$$

(2.59)

and their two Lagrangians are

$$T_m = m\dot{q}_1\dot{q}_2$$

$$U_m = -\frac{\phi_1\phi_2}{m}$$

(2.60)

The above are the most common members of the set of linear energic devices.

Since y_s is a force and $\dot{x}_s$ is a velocity then their product is the power dissipated along coordinate x_s. For a device to be nonenergic it is necessary and sufficient that

$$\sum_s \dot{x}_s y_s = 0 \qquad (2.61)$$

for all coordinates which are needed to describe the behavior of the device. The following theorem due to Duinker gives a sufficient condition on the form of the Lagrangian to assure that the corresponding element is nonenergic.

> **Theorem 2.4**   *If the Lagrangian consists of the product of a generalized velocity multiplied by a function of only the coordinates then the element is nonenergic.*

For a Lagrangian of the form

$$L = \dot{x}_n f(x_1, \ldots, x_n) \qquad (2.62)$$

the corresponding equation $\mathscr{L}_s L = y_s$ for the case when $s = n$ has the form

$$\sum_{k=1}^{n} \frac{\partial f}{\partial x_k} \dot{x}_k - \dot{x}_n \frac{\partial f}{\partial x_n} = y_s \qquad (2.63)$$

and for $s \neq n$ it has the form

$$-\dot{x}_n \frac{\partial f}{\partial x_s} = y_s \qquad (2.64)$$

The total power is given by

$$\sum_{s=1}^{n} \dot{x}_s y_s = \dot{x}_n \left[\sum_{k=1}^{n-1} \frac{\partial f}{\partial x_k} \dot{x}_k \right] - \sum_{s=1}^{n-1} \left[\dot{x}_s \dot{x}_n \frac{\partial f}{\partial x_s} \right]$$

$$= 0 \qquad (2.65)$$

so the element is nonenergic. Two rather obvious corollaries follow.

> **Corollary 2.4.1**   *An element is nonenergic if its Lagrangian is the sum of only such terms as $\dot{x}_i f_i$.*

The proof follows that of the theorem. By considering one term at a time it is true that

$$\sum_{s=1}^{n} \dot{x}_{sj} y_{sj} = 0 \qquad (2.66)$$

and the double sum formed over all terms is also zero.

> **Corollary 2.4.2**   *The Hamiltonian is zero if the Lagrangian is the sum of only such terms as $\dot{x}_j f_j$.*

Using its definition in generalized coordinates the Hamiltonian may quickly be evaluated by the relations

$$H = \sum_s \dot{x}_s \frac{\partial L}{\partial \dot{x}_s} - L$$
$$= \sum_j \dot{x}_j f_j - L$$
$$= 0 \tag{2.67}$$

In mechanical systems an example illustrating the above theorem is the gyroscope. The relationships of the variables needed to describe its basic operation are shown in Figure 2.2. As a result of the transformation of rotating axes the performance of a gyroscope viewed from a stationary reference system is given by

$$\mathbf{T} = \boldsymbol{\omega} \times \mathbf{H} \tag{2.68}$$

where $\mathbf{T}$ is the torque and $\mathbf{H}$ is the angular momentum of the spinning wheel. The product is the vector product. The fact that $\omega_3 = T_3 = 0$ reduces the vector product to the two terms

$$T_1 = \omega_2 H_3$$
$$T_2 = -\omega_1 H_3 \tag{2.69}$$

Both of these can be derived from the Lagrangian

$$L = H_3 \dot{\theta}_1 \theta_2 \tag{2.70}$$

Since the generalized coordinates are angles the generalized forces are torques. This Lagrangian has the form described in Theorem 2.4 so the

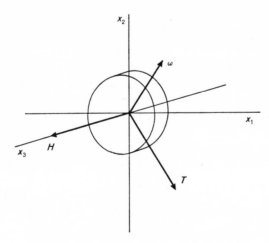

Figure 2.2   The basic variables for a gyroscope.

device is nonenergic. The Hamiltonian for the gyroscopic action is zero. These properties are due to the rectangular relationships of the forces and their displacements. Terms in a Lagrangian which involve the product of a coordinate and a velocity are called *gyroscopic terms*.

It is somewhat astonishing to learn that there are four simple electrical network elements that are derived from gyroscopic terms. A Lagrangian $L = aq_1\dot{q}_1$ defines the situation $v_1 = 0$. This is provided by what is known as a *closed port* wherein the connection of the two terminals prohibits the development of a potential but permits any current to flow. The Lagrangian $L = a\dot{\phi}_1\phi_1$ defines the dual situation $i_1 = 0$ which is provided by what is called an *open port*. Its properties are quite evident by analogy.

The electrical transformer consists of a set of coupled coils and it might be mistaken for an energic device. However, the magnetic circuits and physical arrangements are designed so that essentially the same flux flows through both coils. The induced potentials are therefore proportional to the ratio of the number of turns in the coils. In a well designed transformer the current required to magnetize the primary is small and the currents in the coils are essentially inversely proportional to the turns ratio. These properties are summarized in the equations

$$v_2 = nv_1$$
$$i_1 = -ni_2$$
(2.71)

These characteristics are derived from either of the two Lagrangians

$$L = n\phi_1\dot{q}_2$$
$$L = -\frac{q_1\dot{\phi}_2}{n}$$
(2.72)

These are gyroscopic terms and the transformer quite obviously is a transmissive device.

There is an electrical network element which corresponds to a gyroscope. It is called a *gyrator* and, like the transformer, it is a four-terminal or two-port device. It has the property that a current flow at one port creates a potential at the other. It is described by the equations

$$v_1 = -gi_2$$
$$v_2 = gi_1$$
(2.73)

These can be generated by the Lagrangians

$$L = gq_1\dot{q}_2$$
$$L = -\frac{\phi_1\dot{\phi}_2}{g}$$
(2.74)

If a capacitor is connected to a gyrator the describing equations are

$$v_1 = -gi_2 = gc\frac{dv_2}{dt} = g^2c\frac{di_1}{dt} \tag{2.75}$$

and the input port appears as an inductor. The gyrator also converts an inductor into a capacitor. When two gyrators are connected in tandem the result is an ideal transformer. But two transformers in tandem do not produce a gyrator. From this theoretical point of view the gyrator appears to be possibly more basic and certainly more flexible than the transformer. Adding the gyrator to the list of linear elements permits the deletion of the transformer and one reactor. Because of these unusual properties it is interesting to explore the possibility of realizing the device.

No simple device behaves like a gyrator. A fairly good approximation can be made by applying the Hall effect to a semiconductor. A magnetic field is applied in one direction and terminal pairs are established in each of the other two directions. Since the electromagnetic force on the electron is also expressed as a vector product the analogy with a gyroscope is quite evident.

It the output circuit is open and a current passes through the input circuit then there is a migration of charge normal to both the magnetic field and the current. The device behaves somewhat like a capacitor and there is developed a transverse potential which can be written as $v_2 = R_{21}\dot{q}_1$. Reversing the input and output circuits produces a transverse potential given by $v_1 = -R_{12}\dot{q}_2$ where the negative sign is due to the direction of the electromagnetic force. Assuming that the capacitive effects in the input and output circuits are linear then the potential energy functions have the form

$$\begin{aligned} -U_1 &= \tfrac{1}{2}v_1q_1 = \tfrac{1}{2}R_{12}\dot{q}_2q_1 \\ U_2 &= \tfrac{1}{2}v_2q_2 = \tfrac{1}{2}R_{21}\dot{q}_1q_2 \end{aligned} \tag{2.76}$$

and the Lagrangian is given by

$$L = \tfrac{1}{2}(R_{12}q_1\dot{q}_2 - R_{21}\dot{q}_1q_2) \tag{2.77}$$

There is appreciable resistance in each path. The appropriate dissipation function can be written as

$$G = \tfrac{1}{2}(R_{11}\dot{q}_1{}^2 + R_{22}\dot{q}_2{}^2) \tag{2.78}$$

Consequently the Lagrange equations for the gyrator are

$$\frac{d}{dt}(-\tfrac{1}{2}R_{21}q_2) - (\tfrac{1}{2}R_{12}\dot{q}_2) = v_1 - R_{11}\dot{q}_1$$
$$\frac{d}{dt}(\tfrac{1}{2}R_{12}q_1) + (\tfrac{1}{2}R_{21}\dot{q}_1) = v_2 - R_{22}\dot{q}_2 \tag{2.79}$$

These easily reduce to the simpler form

$$R_{11}i_1 - \bar{R}_{21}i_2 = v_1$$
$$\bar{R}_{21}i_1 + R_{22}i_2 = v_2 \tag{2.80}$$

where $\bar{R}_{21}$ is the average of R_{12} and R_{21}. These are the equations of an ideal gyrator with the unfortunate addition of dissipative terms.

The above constitutes a reasonably complete summary of linear network elements. Nonlinear elements have more complicated Lagrangians but some forms are forbidden. A product of ϕ_j and q_j or their derivatives would imply that the behavior at one port depends on two independent variables. Such activity is unknown in simple elements and is excluded from this discussion. Also, terms such as ϕ^n or q^n are forbidden for negative or fractional exponents. In the first case the Lagrangian would be infinite whenever $\phi = 0$ or $q = 0$. In the second case it might be imaginary for a negative ϕ or q. Sometimes the Lagrangian for a nonlinear element must be indicated as an integral. For example, a nonlinear inductor can not be represented as simply as $\phi = li$ and in general its stored energy is given by

$$U_l = \int_0^t iv \, dt = \int_{\phi_1}^{\phi_2} i \, d\phi \tag{2.81}$$

The stored energy in a nonlinear capacitor is found by the similar relation

$$U_c = \int_0^t iv \, dt = \int_{q_1}^{q_2} v \, dq \tag{2.82}$$

and if two coils are coupled through a nonlinear medium then the total stored energy can be represented as

$$U_m = \int_0^t (i_1v_1 + i_2v_2) \, dt$$
$$= \int_{\phi_{11}}^{\phi_{12}} i_1 \, d\phi_1 + \int_{\phi_{21}}^{\phi_{22}} i_2 \, d\phi_2 \tag{2.83}$$

COMPLEMENTARY FUNCTIONS

Most of the discussion in the remainder of this chapter is limited to electrical networks. There are several reasons. First, in nonrelativistic mechanics the mass is constant. In electrical networks both the inductance and the capacitance may be nonlinear so the electrical network offers a richer variety of situations to explore. Second, a significant portion of modern developments in the theory have been made by workers in electrical engineering so most of the literature deals with electrical problems. This is due in a large part to the fact that nonlinear situations in mechanical systems usually

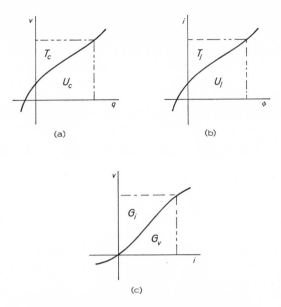

Figure 2.3   The complementary functions for electrical network elements. (a) Capacitor, (b) inductor, and (c) resistor.

are not intentional and their properties are of no interest. On the contrary, electrical and electronic devices and systems are often provided with non-linear characteristics intentionally so as to accomplish a particular objective.

In two companion reports published in 1951 Cherry and Millar showed that the use of the Lagrange method in nonlinear electrical networks requires the introduction of some new functions. These are shown in Figure 2.3. For a capacitor the functions are

$$U_c = \int_0^q v \, dq$$

$$T_c = \int_0^v q \, dv$$

$$(2.84)$$

Although both integrals have the dimension of energy, if the characteristic of the element is nonlinear then only U_c is the energy stored in the element. The complementary function T_c is called the *coenergy*. It has no direct physical significance but it may be necessary for a proper construction of the Lagrangian.

Since these and the following functions describe the state of a network element they are sometimes called *state functions*. The conditions under which these functions exist are described later. Since these functions are components of a Lagrangian it is to be expected that they display some form of stationary properties. This also is demonstrated later in a more general case.

The fact that one integral requires a knowledge of $v(q)$ and the other a knowledge of $q(v)$ introduces the problem of the *inversion* of a function. If a function is linear its inversion is simple. When a function is nonlinear a necessary and sufficient condition for inversion is that each variable is defined uniquely for all values of the other variable. Thus, oscillatory or asymptotic functions are excluded. As a sufficient condition it is adequate to require that both derivatives are never negative, that they are zero in only infinitesimal regions, and that both variables approach infinity jointly. (In the multivariable case the Jacobian determinant of the derivatives must be used.)

For an inductor the complementary functions are

$$U_l = \int_0^\phi i \, d\phi$$
$$T_l = \int_0^i \phi \, di$$

(2.85)

where U_l is the energy and T_l is the coenergy. For a resistor the dissipation function is divided into the two components

$$G_v = \int_0^i v \, di$$
$$G_i = \int_0^v i \, dv$$

(2.86)

These have the dimensions of power. Millar has given to G_v the name *content* and the function G_i is called the *cocontent*. Figure 2.3(c) demonstrates the relationship

$$G_v + G_i = iv$$

(2.87)

for a simple case where the characteristic curve passes through the origin. This indicates a *passive* device. In the more general situation this relationship between the complementary functions is described by what is called the *Legendre transformation*. This is shown in Figure 2.4. Regardless of the location of the initial point it is true that

$$\int_{x_0}^x y \, dx = B$$
$$\int_{y_0}^y x \, dy = C$$

(2.88)

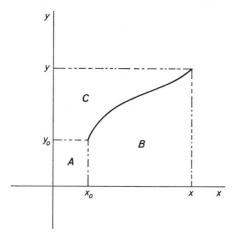

Figure 2.4 Development of the Legendre transformation.

Under the condition that either $x_0 = 0$ or $y_0 = 0$ then $A = 0$ and it is always true that

$$\int_{x_0}^{x} y \, dx + \int_{y_0}^{y} x \, dy = xy \tag{2.89}$$

This is the Legendre transformation. It is equivalent to an integration by parts and is also valid if the initial point corresponds to negative values for x or y provided that the other variable is zero. This transformation applies to the dissipation functions whether the element is active or passive and it applies to the energy functions of devices both with and without hysteresis.

If q is taken as the generalized coordinate and i is the corresponding velocity then the Lagrangian is constructed in the form

$$L = \int_{0}^{i} \phi \, di - \int_{0}^{q} v \, dq$$

$$= T_l - U_c \tag{2.90}$$

and for the dissipation function the content must be used to provide the proper variable of integration. The Lagrange equation then becomes

$$\frac{d}{dt} \frac{\partial}{\partial i} \int_{0}^{i} \phi \, di + \frac{\partial}{\partial q} \int_{0}^{q} v \, dq = v - \frac{\partial G_v}{\partial i} \tag{2.91}$$

which reduces to

$$\frac{d\phi}{dt} + v_c = v - \frac{\partial G_v}{\partial i} \tag{2.92}$$

This is clearly a sum of voltages associated with the charge q and its current i. The situation may be visualized as a series circuit containing the three different devices plus a voltage source. However, there is at this time no logical basis for such an assumption. The laws governing the flow of current and the distribution of voltage in a network are presented in the following section and there the methods for selecting generalized velocities and their relation to the remainder of the network are discussed in considerable detail.

If ϕ is taken as the generalized coordinate and v is its corresponding velocity then the Lagrangian is constructed in the form

$$L = \int_0^v q \, dv - \int_0^\phi i \, d\phi$$
$$= T_c - U_l \tag{2.93}$$

and in this case the cocontent is used for the dissipation function. The Lagrange equation then becomes

$$\frac{d}{dt} \frac{\partial}{\partial v} \int_0^v q \, dv + \frac{\partial}{\partial \phi} \int_0^\phi i \, d\phi = i - \frac{\partial G_i}{\partial v} \tag{2.94}$$

which reduces to

$$\frac{dq}{dt} + i_l = i - \frac{\partial G_i}{\partial v} \tag{2.95}$$

This represents a sum of currents related to the flux ϕ and its voltage v. The network can be visualized as a parallel circuit of the three elements plus a current source. The voltage is that which appears across all of the elements and the currents are those which flow into a junction. Again this is a pure presumption and the results of the next section must be consulted to justify such a supposition. The following theorem due to Duinker involves the complementary functions.

Theorem 2.5   *In a network of nonlinear reactances but linear resistances the work expended by transients in bringing the network from a rest state to a steady state equals the excess of the sum of the final electric energy and coenergy over the sum of the final magnetic energy and coenergy.*

Using the charges q_k as coordinates the Lagrangian is

$$L = \sum_k T_{lk}(\dot{q}) - \sum_k U_{ck}(q) = T_l(\dot{q}) - U_c(q) \tag{2.96}$$

where the notation indicates that T_{lk} and U_{ck} are functions of various $\dot{q}$ and q respectively. Under steady-state conditions all time derivatives are zero so the Lagrange equations are reduced to

$$-\frac{\partial L}{\partial q_k} = v_{ck} = v_k - R_k i_k \tag{2.97}$$

Under the final conditions the sum of the electric energy and coenergy is given by

$$T_c + U_c = \sum_k q_k v_{ck} = \sum_k q_k(v_k - R_k i_k) \tag{2.98}$$

Assuming that each impressed voltage v_k is constant then the total energy delivered by all of the sources is

$$W_s = \sum_k q_k v_k \tag{2.99}$$

and a direct substitution gives

$$T_c + U_c = W_s - \sum_k R_k q_k \dot{q}_k \tag{2.100}$$

From the definition of the magnetic complementary functions it is true that

$$\phi_k = \frac{\partial L}{\partial \dot{q}_k} \tag{2.101}$$

so the sum of the magnetic energy and coenergy can be written in the various forms

$$
\begin{aligned}
T_l + U_l &= \sum_k \dot{q}_k \phi_k \\
&= \sum_k \dot{q}_k \int_0^t \frac{d}{dt} \frac{\partial L}{\partial \dot{q}_k} \, dt \\
&= \sum_k \dot{q}_k \int_0^t \left(v_k - R_k i_k + \frac{\partial L}{\partial q_k} \right) dt
\end{aligned}
\tag{2.102}
$$

For any general circuit which contains steady-state magnetic energy and coenergy $\dot{q}_k$ can not be zero and the circuit can not contain a capacitor so the above reduces to

$$T_l + U_l = W_f - \sum_k R_k \dot{q}_k q_k \tag{2.103}$$

where W_f is the energy which would have been expended if the final value of each $\dot{q}_k$ had existed all of the time. The difference between this ultimate energy and the energy supplied by the sources equals the energy dissipated in transients. Subtracting Equation 2.103 from Equation 2.100 gives the final result

$$(T_c + U_c) - (T_l + U_l) = W_t \tag{2.104}$$

where W_t is the energy lost in transients.

NETWORK EQUATIONS

If the Lagrange, or indeed any other technique, is to be applied to a complex network then certainly the feasibility and efficacy of its application depend on the properties of the network, Some topological characteristics of networks are discussed in the first chapter. A few are repeated here and some new concepts are introduced.

The network is assumed to consist of a collection of branches each of which contains one type of electrical device. The branches are joined at each end to other branches at a *node*. To each branch there are assigned two variables. One is the current passing through the branch and the other is the voltage across the branch. These are called the *branch variables*. Their absolute polarity must frequently be assumed but their relative polarity is shown in Figure 2.5. Part (a) shows the polarity for branches within the network and part (b) shows the polarity normally assigned at the terminals. It is further assumed that the devices in each branch are known and that their properties are uniquely defined. In each branch the known functional relationship between the branch variables is called the *branch relation*.

A chain of branches which returns to the initial node is called a *circuit* or a *mesh*. A set of branches which does not contain a mesh is called a *tree*. If the tree spans all of the nodes it is called a *maximal tree*. Figure 2.6 shows a maximal tree. The dotted lines are the remaining branches of the network. They are called *links* or *chords*. If a network contains two vertices which do not lie in a circuit then the network is said to be *separable*. Figure 2.7 shows a separable network. In all of the following discussion such networks are specifically excluded. If a chain of branches connects each pair of nodes then the network is said to be *connected*.

Lemma 2.1   *In a connected network it is always possible to construct a maximal tree.*

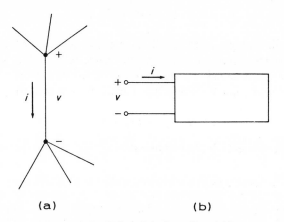

(a)   (b)

Figure 2.5   Polarity of the branch variables.

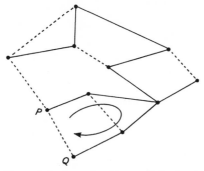

Figure 2.6   A maximal tree and its chords.

This is established by giving a general method for the construction. Since the network has a finite number of branches they can be enumerated in an arbitrary fashion. Each branch can then be examined in order. If it lies in any mesh of the network it is deleted. This removal can not isolate a node or disconnect the network. After the last branch is tested the result is a maximal tree. Sometimes a different enumeration produces a different maximal tree. The set of all permutations produces all possible maximal trees.

Lemma 2.2   *A given branch of a network is the branch of at least one maximal tree and the chord of at least one other maximal tree.*

If the given branch is the chord of a maximal tree then when it is added to the tree one and only one mesh is formed. This contains at least one other branch of the tree. When this branch is removed another maximal tree is formed and the given branch is a branch of this tree. This proves the first part of the lemma. If the given branch is originally a branch of a maximal tree then, since the graph is nonseparable, its nodes must lie in a mesh. This mesh can be formed by adding a chord to the tree. If the given branch is removed then a maximal tree is produced and the given branch is a chord of this tree. This establishes the second part of the lemma.

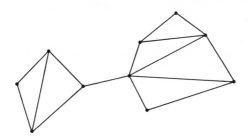

Figure 2.7   A separable network.

Reference to Figure 2.6 shows that if there are N nodes in the network then the maximal tree has $N - 1$ branches. This is easily established by systematically disassembling the tree and observing that there is one node for each branch except the last which has two. It is also important to observe that if there are B branches and $N - 1$ maximal tree branches then there must be $B - N + 1$ chords. This is a significant fact and the number $B - N + 1$ is called the *circuit rank* for the network.

It is a direct consequence of the method of construction of a maximal tree that there exists one and only one mesh which is properly associated with each chord. It is the mesh which is opened when the chord is deleted. Each of these meshes is called a *basic mesh*. Figure 2.6 shows the basic mesh corresponding to the chord between nodes P and Q. It shows a completely artificial and fictitious current which is considered to be flowing around this mesh. Such currents are called *mesh currents* and they play an important role in network theory.

In a classic report published in 1847 the young Kirchhoff formulated the two most important laws of network theory. They are stated below as postulates. A translation of his report is listed in the bibliography and is recommended for supplementary reading. In this report Kirchhoff also developed the concepts of the circuit rank and the maximal tree of a network. In 1851 Helmholz introduced circuital currents such as those which are called mesh currents here and demonstrated how they can be used in conjunction with Kirchhoff's laws to solve network problems. Since that time topology has developed as an extremely rigorous branch of mathematics. In 1944 Ingram and Cramlet applied these newer methods and ideas to electrical network theory and produced a more precise logical structure describing network analysis. Synge and Saltzer later presented a more palatable treatment of some of the material and made additional contributions. Significant reports by these workers are listed as references for this chapter. This section reviews the more important portions of these developments. Following are the laws of Kirchhoff.

Postulate 2.1   *The sum of the branch voltages around a mesh is zero.*

Postulate 2.2   *The sum of the branch currents at a node is zero.*

Appropriate algebraic signs must be attached to the terms in each sum. The first law assumes that there is no vortex motion in the force field from which the potentials are derived. The second law assumes that the node has no capability of storage. The following lemma is an immediate extension of Kirchhoff's current law.

Lemma 2.3   *The total current flowing through any continuous surface intersecting the branches of a network is zero.*

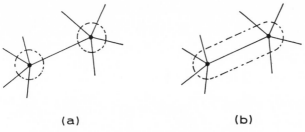

(a)   (b)

Figure 2.8   An extension of Kirchhoff's current law.

Part (a) of Figure 2.8 shows a planar view of two nodes of a network surrounded by two such continuous surfaces. The total current flowing through each surface is zero by Postulate 2.2. Part (b) of Figure 2.8 shows one surface surrounding both nodes. The total current flowing through this surface includes the same branch currents which flow through the two surfaces of part (a) except for the current in the branch connecting the two nodes. In the original sums this current has equal and opposite values so its exclusion in forming the sum for the surface in part (b) has no effect. This is true if several branches connect the nodes. The process can obviously be continued to form a surface enclosing any set of connected nodes and the sum of the currents remains zero.

From geometric considerations it is true that the set of branches intersected by such a continuous surface as that described above constitute in each case the minimal set which divides the network into two disconnected parts. Such a set of branches is called a *cut-set*.

Corollary 2.3.1   *The sum of the branch currents in a cut-set is zero.*

This is merely an alternate statement of Lemma 2.3.

Lemma 2.4   *Any set of mesh currents produce branch currents which satisfy Kirchhoff's current law at every node.*

Since each mesh current leaves every node which it enters then it must be true at every node that

$$\sum_m i_m = 0 \qquad (2.105)$$

where each i_m is a mesh current. In accordance with the particular manner in which the mesh currents are related to the branch currents at each node all of these terms can be rearranged to give all of the branch currents so it is also true at every node that

$$\sum_b i_b = 0 \qquad (2.106)$$

where each i_b is a branch current.

Lemma 2.5   The total current flowing into a maximal tree is zero.

This is easily established by assuming that there is flowing any permissible set of chord currents which are not all zero. These are the only sources of current for the tree and they make equal and opposite contributions at the two ends of each chord so the total flow into the tree is zero.

Lemma 2.6   The set of chord currents uniquely determines the tree branch currents.

Since the tree has branches but no meshes it must consist of a set of terminal branches plus a set of interior branches. To each terminal branch is connected a chain of interior branches which leads either to a node to which several branches are connected or to the other terminal branch. In the first and more complex case the terminal branches can be enumerated and the tree can be disassembled in an orderly fashion. Starting with the first terminal branch one deletes it plus the chain of connected interior branches until a node is reached where several branches join. The operation stops at this point. A continuation of this process through the succession of terminal branches ultimately leads to a single chain of interior branches joining the last two terminal branches.

During the above disassembly it is possible to use Kirchhoff's current law to find the current in each branch as it is deleted. The chord currents determine the terminal branch currents uniquely and these determine the currents uniquely in the chain of connected internal branches. When the tree is reduced to the last two terminal branches one can start at either end and delete branches until the other end is reached. At each node there is only one branch with undetermined current and Kirchhoff's current law is sufficient for the determination. On reaching the last terminal branch the flow at one end must be the negative of the flow at the other end. This is consistent with Lemma 2.5 so the process cannot lead to a contradiction.

Lemma 2.7   If a network contains N nodes then Kirchhoff's current law provides $N - 1$ independent equations.

A tree has the unique property that the deletion of any one of its branches divides it into two parts. Since a maximal tree spans every node of the graph the deletion of one of its branches partitions the nodes of the graph into two disjoint sets. Therefore for a given branch of the tree it is always possible to construct a cut-set which contains the given branch but no other tree branch. All other branches in the cut-set are chords. Such a cut-set is called a *basic cut-set* and some properties of these are discussed in the first chapter. Figure 2.9 shows a graph with one of its maximal trees and three basic cut-sets.

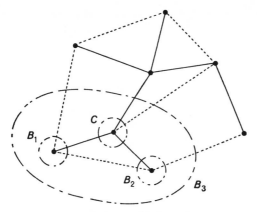

Figure 2.9 Dependent and independent cut-sets.

These are B_1, B_2, and B_3. Since there are $N - 1$ branches on a maximal tree there are $N - 1$ basic cut-sets. By definition a basic cut-set meets the geometric requirement of Lemma 2.3 and each contains a unique branch current. Hence the set of basic cut-sets generates $N - 1$ independent linear algebraic equations.

It must also be shown that no more than $N - 1$ independent relations are possible. Figure 2.9 shows a cut-set C which corresponds to the application of Kirchhoff's current law at another node. This creates the sum of three tree branch currents and one chord current. Each of the tree branch currents can be expressed as a linear combination of the variables appearing in the three original basic cut-sets. The chord current in C can be expressed similarly in terms of the variables appearing in B_3. Therefore the sum of the currents created by cut-set C involves the same variables as those in the three basic cut-sets. Since all of the relations are linear the information must be either incompatible or redundant. Application of the method which established Lemma 2.3 shows that it must be redundant. Hence in a network with N nodes there are exactly $N - 1$ independent equations involving the branch currents.

Consequently, in the above example it is possible to delete basic cut-set B_3 and use cut-set C instead. It is observed that C introduces the same independent variable as B_3. This is the current in the tree branch which is common to each. A cut-set C which is not basic for a given maximal tree but which introduces a variable which is not found in the other cut-sets is called an *independent cut-set*.

Corollary 2.7.1   *The number of independent currents in a network equals the number of chords.*

There are altogether B different branch currents constrained by $N - 1$ equations. There remain $B - N + 1$ currents which can be selected arbitrarily. This is the number of chords.

Lemma 2.8   *For each maximal tree there exists a set of mesh currents which generate a given set of branch currents.*

The configuration of the tree determines a unique set of basic meshes. The current in each of these meshes is equal to the current in its chord. Lemma 2.6 shows that the tree branch currents are determined uniquely and, following Kirchhoff's current law, these must correspond to the given set.

Corollary 2.8.1   *No current flows in a network which is its own maximal tree.*

This is established by adding chords to the network and repeating the previous argument with vanishing chord currents. The result is equivalent to a network which is its own maximal tree.

Lemma 2.9   *Every branch in a nonseparable network passes at least one chord current.*

The given branch originally was in at least one mesh since the network is nonseparable. When a maximal tree is constructed this mesh is destroyed either by deleting the given branch or another in the mesh. In either case the given branch lies in a basic mesh uniquely associated with a chord.

Lemma 2.10   *Any set of node voltages produces branch voltages that satisfy Kirchhoff's voltage law around every mesh.*

Selecting any mesh in the network and traversing it in the assigned direction produces a set of branch voltages derived uniquely from the node voltages. This sequence of numerical differences must have a zero sum since the initial and terminal numbers are the same.

Lemma 2.11   *The branch voltages of a maximal tree uniquely determine the chord voltages.*

To each chord there corresponds a basic mesh which, except for the chord, is composed entirely of branches of the maximal tree. Therefore Kirchhoff's voltage law is sufficient to evaluate uniquely each chord voltage.

Lemma 2.12   *For each maximal tree there exists a set of node voltages which generate a given set of branch voltages.*

The voltage at one node can be assigned arbitrarily. Then one can proceed from this point taking the tree branches in order and using the given branch voltages to compute the node voltages. When this is done the previous lemma shows that the chord voltages must also be consistent with the given set. The method also shows that there are infinitely many sets of node voltages which give the required branch voltages.

Lemma 2.13 *If a network contains B branches and N nodes then Kirchhoff's voltage law provides B − N + 1 independent equations.*

Since each chord appears uniquely in its own basic mesh there must be at least as many independent equations as chords. To prove that there can be no more than $B - N + 1$ independent relations it is necessary to show that any other mesh in the network produces an equation which is a linear combination of the relations produced by the basic meshes. From the discussion of Lemma 2.9 it is apparent that each branch of the network lies on at least one basic mesh. The branch voltage can be computed by solving the Kirchhoff voltage equation around the basic mesh. All branch voltages can be expressed in this manner. When applying Kirchhoff's voltage law to a mesh other than the basic meshes the resultant equation is only a linear combination of the equations derived from the basic meshes and is not an independent relation. In a manner analogous to the use of tree branches as a basis for constructing independent cut-sets the chords form a basis for constructing independent voltage sums. And analogous to independent cut-sets for a given maximal tree there are also *independent meshes*.

Corollary 2.13.1 *The number of independent voltages in a network equals the number of branches on a maximal tree.*

There are altogether B different branch voltages constrained by $B - N + 1$ equations. There remain $N - 1$ voltages which can be selected arbitrarily. This is the number of branches on a maximal tree.

The above discussion summarizes the more fundamental results of imposing Kirchhoff's laws on a network. They form an impressive logical structure but it is somewhat confusing to derive from them a systematic attack on a network problem. The concept of what is called a *complete set* of network variables facilitates analysis. Such a complete set has the following properties:

1. The variables are independent relative to both Kirchhoff's laws and the branch relations.

2. An appropriate linear combination of the variables in accordance with only Kirchhoff's laws completely determines one branch variable in each branch.

The first requirement assures that no member of the set is redundant. The second requirement guarantees enough information to produce a solution. It is quite apparent that the best collection of independent variables for solving any problem is a complete set. The previous discussion has established the following lemmas.

Lemma 2.14   *The branch voltages of a maximal tree form a complete set of network variables.*

Lemma 2.15   *The chord currents of a maximal tree form a complete set of network variables.*

In traditional network analysis the above two lemmas indicate methods of solution known as *node analysis* and *mesh analysis* respectively. Many examples of these methods appear in standard electrical engineering texts so no illustrations are given here. However, at the end of this section one network is analyzed in four different manners including these two basic methods.

In a significant report listed in the bibliography Brayton and Moser introduced a form of analysis using mixed variables. The graph of the network is divided arbitrarily into two components, a subgraph and its complement. A typical situation is shown in Figure 2.10. A maximal tree of the graph is then constructed so that its branches which lie in the subgraph form a maximal tree therein. This uniquely separates the subgraph into tree branches and chords. Its complement is also divided uniquely into tree branches and chords but it cannot be assumed that the branches form a maximal tree since they may not be connected.

Lemma 2.16   *The branch voltages of a maximal tree in a subgraph plus the chord currents in its complement form a complete set of network variables.*

This is established by observing first that the tree branch voltages are independent and they determine all voltages in the subgraph. Since the tree in

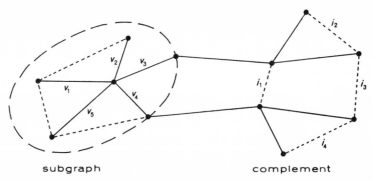

subgraph   complement

Figure 2.10   Typical graph for hybrid analysis.

the subgraph is maximal then the mesh currents generated by its chords must lie entirely within the subgraph. Therefore the branch currents in the complement depend on only the mesh currents generated by chord currents in the complement. These are independent since they are chord currents of the original graph. The stipulated set of variables form a complete set since they are independent and since they determine branch voltages in the subgraph and branch currents in its complement. Since the subgraph and the maximal tree can be selected in many ways this method provides great variety for selecting the complete set of network variables. In any case the generation of independent current and voltage sums can be assured by using a basis which is a *hybrid basis* consisting of the tree branches in the subgraph plus the chords in the complement.

When applying the Lagrange method to a hybrid set of network variables the total dissipation function has the form

$$G_{iv} = \sum_s \int_0^{v_s} i_s \, dv_s + \sum_c \int_0^{i_c} v_c \, di_c \tag{2.107}$$

This is called the *hybrid dissipation function*. For the sum in the resistive branches of the subgraph the integrals use the branch voltages as the independent variables. In the complement the branch currents are used. These may be functions of several of the chord currents in the complete set of variables. The above choice of variables is not arbitrary. It is necessary to provide the proper functional relationship for the Lagrange formulation.

An investigation of the application of hybrid analysis shows that the same network can produce sets of differential equations of widely different forms. This suggests finding a procedure which produces the simplest forms. In 1959 Bryant published a report which makes a very penetrating investigation into the true complexity of a network. His method produces equations of the simplest form but his analysis concentrates on linear networks. In 1966 Stern extended his method to nonlinear networks. Although the important aspects of the method are outlined here the original reports are listed in the bibliography and are strongly recommended for collateral reading.

The first step in the method is to define a reduced set of network variables. Paraphrasing Bryant these will be called a *dynamic set* of network variables. Such a set has the following properties:

1. The variables are independent relative to Kirchhoff's laws and the branch relations.

2. The values of the variables are sufficient to determine the state of the network through purely algebraic operations.

The difference between a complete set and a dynamic set of network variables is shown in Figure 2.11. In part (a) complete sets consist of either v and v_r,

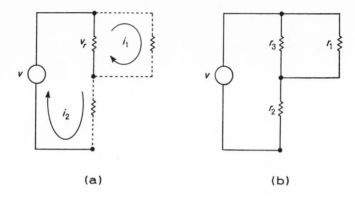

(a)   **(b)**

Figure 2.11 Sets of network variables. (a) Complete set and (b) dynamic set.

i_1 and i_2, or a hybrid set. If these are used as independent variables in the network equations then their evaluation immediately establishes what is happening in each branch. In part (b) the dynamic set consists entirely of the voltage v. Since it is assumed that the resistance of each resistor is known the specification of the voltage v is sufficient to determine the state of the network by algebraic operations. However, to calculate what is happening in each branch it is necessary to introduce a complete set of variables either explicitly or implicitly.

The second step in the method involves a special technique for selecting the maximal tree. The procedure has the following steps:

> **1.** The resistors and inductors are replaced with open circuits. Any capacitive meshes which remain are opened by deleting one arbitrary capacitor from each mesh. This produces a set of capacitive trees. Such a set is called a *forest* and is designated F_c.
>
> **2.** The resistors and capacitors are replaced with short circuits. Inductors are deleted arbitrarily to form a maximal tree in the reduced network. This produces a forest F_l in the original network.
>
> **3.** Resistors are added arbitrarily to F_c and F_l to produce a maximal tree in the original network.

The last step is always possible since neither of the first two operations isolate a node. It should be observed that the above procedure does not produce a unique maximal tree. Numerous trees are usually possible but all produce equations of equivalent complexity. The following lemma occupies a central position in this method.

Lemma 2.17   *The voltages across the capacitors in the branches of F_c and the currents through the inductors in the chords of F_l form a dynamic set of network variables.*

To prove this lemma it is necessary to establish the fact that the given variables either immediately determine a branch variable in each branch or provide sufficient information that such determination can be made by algebraic operations. The discussion will treat capacitors, inductors, and resistors in that order.

As a consequence of the special method for constructing the maximal tree the basic meshes for the capacitive chords contain only capacitors. If they contained resistors or inductors then during the construction of the tree there would be no circuits and the capacitive chords would not be chords. Therefore Lemma 2.9 applies to the capacitive portion of the network and a simple mesh analysis evaluates the voltages across all remaining capacitors in the network. This analysis does not use the branch relations and nonlinear characteristics have no effect on the solution.

Corresponding to the inductive tree which was constructed with all resistors and capacitors shorted is a set of independent cut-sets. These contain only inductive branches. They also divide the network into two parts when the short circuits are replaced with the original elements. Therefore they provide a set of independent equations which evaluate all of the inductive branch currents as functions of the inductive chord currents. Again only linear relations are used. Figure 2.12(a) shows a network with various types of meshes. It shows two inductive cut-sets and one capacitive mesh.

The capacitors can now be considered as voltage sources and the inductors can be represented as current sources. The network then appears as shown in Figure 2.12(b). A systematic method of solution is to use the maximal tree as a basis for a hybrid analysis. The inductive branches are placed in the subgraph and the capacitive branches are in the complement. In Figure 2.12(b) the subgraph lies inside the dashed line. The resistive branches are assigned arbitrarily to the part which appears more appropriate. Then the voltages of the resistive tree branches in the subgraph and the currents of the resistive chords in the complement are added to the dynamic set of variables to produce a complete set.

Node analysis is used in the subgraph and mesh analysis is used in the complement. If in the subgraph there are N_1 nodes then there are $N_1 - 1$ tree branches and as many independent cut-sets. These are sufficient to determine all node potentials and hence all resistive branch voltages. These appear as functions of the inductor currents.

If in the complement there are B_2 branches and N_2 nodes then there are $B_2 - N_2 + 1$ chords and an equal number of independent mesh equations.

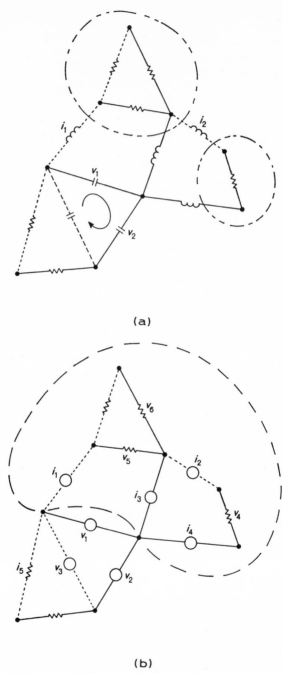

(a)

(b)

Figure 2.12 Special hybrid analysis. (a) The inductive cut-sets and the capacitive mesh and (b) the resistive network.

These are sufficient to determine the currents in all resistive chords and hence all resistive branches. These currents are functions of the capacitor voltages.

In the solutions for the resistive branch variables the branch relations are used. Since these may be nonlinear functions the existence and uniqueness of the solutions depends on the nature of the functions which are involved. Assuming that suitable solutions exist then the above argument establishes the lemma since at least one variable in each branch can be derived from the original set by purely algebraic operations. In particular it is true that the resistive chord currents and resistive tree branch voltages can be expressed in the form

$$i_{cr} = i_{cr}(v_{tc}, i_{cl})$$
$$v_{tr} = v_{tr}(v_{tc}, i_{cl})$$

(2.108)

where the notation signifies that the resistive currents and voltages can be written as functions of various combinations of capacitive tree branch voltages and inductive chord currents. The first subscript indicates location and the second indicates the type of element.

A mesh of the general form shown in part (a) of Figure 2.13 gives an equation such as

$$\dot{\phi}_{cl} + v_{tc}(q_{tc}) + \dot{\phi}_{tl} + v_{tr}(v_{tc}, i_{cl}) = 0$$

(2.109)

where v_{tc}, $\dot{\phi}_{tl}$, and v_{tr} may represent several elements of their type in the mesh. The form of Equation 2.109 suggests defining a new variable $\phi_j = \phi_{cl} + \phi_{tl}$ and the equation becomes

$$\dot{\phi}_j = -v_{tc}(q_{tc}) - v_{tr}(v_{tc}, i_{cl})$$

(2.110)

A cut-set based on a terminal tree branch as shown in part (b) of Figure 2.13 gives an equation in the form

$$\dot{q}_{cc} + i_{cl}(\phi_{cl}) + \dot{q}_{tc} + i_{cr}(v_{tc}, i_{cl}) = 0$$

(2.111)

A more general cut-set may involve i_{tl} and i_{tr} but this introduces no additional variables since such terms are also functions of v_{tc} and i_{cl}. The definition $q_k = q_{cc} + q_{tc}$ reduces the above equation to

$$\dot{q}_k = -i_{cl}(\phi_{cl}) - i_{cr}(v_{tc}, i_{cl})$$

(2.112)

To reduce Equations 2.110 and 2.112 to the simplest form it is necessary to show that the inductive chord currents and capacitive tree branch voltages can be expressed as $i_{cl} = i_{cl}(\phi_j)$ and $v_{tc} = v_{tc}(q_k)$. This is established for the inductive case and the capacitive case is left as an exercise. If there are B_l inductive branches, N_{cl} inductive chords, and N_{tl} inductive tree branches then there are B_l branch relations, N_{cl} equations defining as many ϕ_j, and N_{tl} independent cut-sets. Therefore there are $B_l + N_{cl} + N_{tl}$ independent equations. To express all i_{cl} there are needed N_{cl} equations. Hence $B_l + N_{tl}$

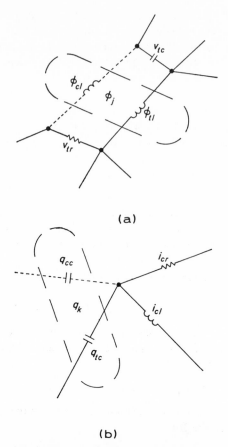

(a)

(b)

Figure 2.13 Generation of Bryant-Stern equations. (a) General mesh and (b) general cut-set.

variables can be eliminated. Since there is one ϕ_j for each basic mesh with inductive chord there are $2B_l + N_{cl}$ variables originally and $B_l + N_{cl} - N_{tl}$ or $2N_{cl}$ variables after the elimination. The result is N_{cl} equations and $2N_{cl}$ variables. With proper elimination of variables the N_{cl} equations can define each i_{cl} and the $2N_{cl}$ variables can include only the i_{cl} and the ϕ_j. A dual argument exists for the v_{tc}. Therefore each i_{cl} and v_{tc} can be expressed as

$$i_{cl} = i_{cl}(\phi_j)$$
$$v_{tc} = v_{tc}(q_k)$$

(2.113)

where the notation represents any combination of members from the sets of all ϕ_j and all q_k. The above elimination of variables involves nonlinear branch relations and the existence and uniqueness of the solutions indicated by Equations 2.113 depend on the behavior of those functions. Substituting Equations 2.113 into Equations 2.110 and 2.112 gives the final set of differential equations

$$\dot{\phi}_j = \dot{\phi}_j(\phi_j, q_k)$$

$$\dot{q}_k = \dot{q}_k(\phi_j, q_k) \tag{2.114}$$

These are the Bryant-Stern equations. A review of the procedure shows that one of these equations is generated only at a node or in a mesh where dissimilar elements are present. The variables are selected such that all of the differential equations are of first order and there is one equation for each variable. When a set of differential equations is arranged in this fashion it is said to be in *normal form*. Not only is the Bryant-Stern procedure significant because it produces fewer and simpler differential equations but also because the equations appear in a standard mathematical form which is the point of departure for most advanced analytical methods.

A network which has no external connections is called a *complete network*. If it satisfies both of Kirchhoff's laws it is called a *Kirchhoffian network*. The following theorem due to Tellegen is one of the most fundamental results in network theory.

Theorem 2.6 *In a complete Kirchhoffian network the sum of the products of the current and the voltage in each branch is zero.*

Let branch b connect nodes j and k which have potentials v_j and v_k. Then it is always true that

$$\sum_b i_b v_b = \frac{1}{2} \sum_j \sum_k (v_j - v_k) i_{jk}$$

$$= \sum_j v_j \sum_k i_{jk} \tag{2.115}$$

If the network satisfies Kirchhoff's current law then

$$\sum_k i_{jk} = 0 \tag{2.116}$$

since this represents the total current flowing out of node j. It is therefore quite apparent that

$$\sum_b i_b v_b = 0 \tag{2.117}$$

Although only Kirchhoff's current law is used explicitly in the proof, the ability to assign unique v_j implies the existence of Kirchhoff's voltage law so the network must be entirely Kirchhoffian. The dual proof involving mesh currents and branch voltages is left as an exercise. This theorem is quite significant because it shows that Kirchhoff's laws are sufficient to establish conservation of energy and power in the network. It is also important to observe that the branch relations are not involved so the theorem applies broadly to any complete network regardless of the properties of the elements in the branches.

Theorem 2.7   *In a complete Kirchhoffian network content is conserved.*

It is true at each node that

$$\sum_b (i_b + di_b) = 0 \tag{2.118}$$

for any variations di_b in the branch currents. Since the original currents form a Kirchhoffian set then it must be true at every node that

$$\sum_b di_b = 0 \tag{2.119}$$

so the displacements in current must also satisfy Kirchhoff's current law. Theorem 2.6 applies to the displacements as well as to the original currents and this gives

$$\sum_b v_b \, di_b = 0 \tag{2.120}$$

so the change in total content throughout the complete network is always zero. Integrating along a Kirchhoffian path for all current variables between states A and B of the network gives

$$\int_A^B \sum_b v_b \, di_b = \sum_b \int_A^B v_b \, di_b = \sum_b \Delta G_{vb} = 0 \tag{2.121}$$

where each ΔG_{vb} is calculated between states A and B in branch b.

A more useful interpretation of Equation 2.121 is found by isolating the branches containing sources. Then the content supplied by the sources equals the negative of the content appearing in the remainder of the network. The same type of interpretation applies to the following theorem which is established by a dual argument.

Theorem 2.8   *In a complete Kirchhoffian network cocontent is conserved.*

Theorems 2.7 and 2.8 are completely general. They apply equally well to active systems by including initial values of content and cocontent. Since the derivations assume no relationships between i_b and v_b the theorems are valid for network elements of any type including those with differential or integral relationships and those with hysteresis.

Theorem 2.9   *In any network the content is stationary against current variations and the cocontent is stationary against voltage variations.*

To be stationary, for example, against current variations means that the currents in the network distribute themselves in such a manner that the function does not change its value for small, arbitrary displacements in the currents provided that the displacements satisfy Kirchhoff's current law. Since the displacements are otherwise arbitrary they may create branch voltages which violate Kirchhoff's voltage law. By Theorems 2.7 and 2.8 the total content and cocontent do not change in a complete network so the above theorem is immediately true in a trivial sense for such a network. The theorem has greater significance when applied to the internal content or cocontent of a network to which current and voltage sources may be connected. The change in internal content, for example, is given by the relations

$$\delta G_v = \sum_b v_b \, \delta i_b$$
$$= \sum_j v_j \sum_k \delta i_{jk} \qquad (2.122)$$

where the sums are formed over only the internal branches. At all internal nodes Kirchhoff's current law constrains the displacements in the currents to a zero sum. At the terminal nodes if there is no change in the external currents then the internal displacements also have a zero sum. Under these conditions

$$\delta G_v = 0 \qquad (2.123)$$

and a dual argument establishes the equation

$$\delta G_i = 0 \qquad (2.124)$$

It should be observed that the content and cocontent are stationary against only current and voltage changes respectively. The total power is not stationary in a nonlinear network as it is in a linear one. This is illustrated by an example. Figure 2.14(a) shows a simple resistive network. One resistor is linear with a resistance of 1 ohm and the other has a cubic characteristic. Assuming for simplicity that $v = 1$ volt then the content, cocontent, and power are given by

$$G_v = \frac{i_1^{\,2}}{2} + \frac{i_2^{\,4}}{4}$$
$$G_i = \frac{i_1^{\,2}}{2} + \frac{3i_2^{\,4}}{4} \qquad (2.125)$$
$$P = i_1^{\,2} + i_2^{\,4}$$

and the total current under equilibrium conditions is 2 amperes. Therefore any assumed displacements in currents must satisfy the relation

$$i_1 + i_2 = 2 \qquad (2.126)$$

Part (b) of Figure 2.14 shows how the content, cocontent, and power vary as i_2 is varied between the permitted extremes. The content has a minimum

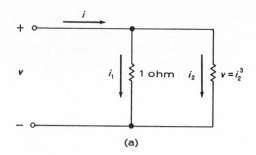

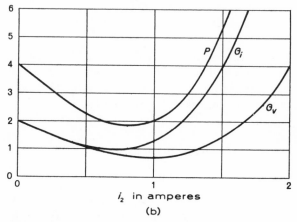

Figure 2.14 (a) A nonlinear network. (b) The content, cocontent, and power.

at the equilibrium point. Both G_i and P have minima also but these do not occur at the value of current which the nonlinear resistor actually passes.

Brayton and Moser have presented a new method for deriving the network equations which is based on the content of the network. Their analysis considers the network as a subgraph of resistors to which inductors and capacitors are connected. Adherence to Kirchhoff's laws is sufficient to assure that Theorem 2.7 applies and Equation 2.121 can be written in the form

$$\int_A^B \sum_b v_{bl}\, di_{bl} + \int_A^B \sum_b v_{bc}\, di_{bc} + \int_A^B \sum_b v_{br}\, di_{br} = 0 \qquad (2.127)$$

where the terms apply to the inductors, capacitors, and resistors respectively. Integrating the second integral by parts changes the variable of integration and gives

$$\int_A^B \sum_b v_{bl}\, di_{bl} - \int_A^B \sum_b i_{bc}\, dv_{bc} + S = 0 \qquad (2.128)$$

where S is given by

$$S = \sum_b i_{bc} v_{bc} \Big|_A^B + \sum_b \int_A^B v_{br}\, di_{br} \qquad (2.129)$$

The function S is called the *mixed potential function*. It depends on only the terminal conditions. Considering state A as constant and state B as variable then it is defined only to a constant which depends on the initial conditions. From Equation 2.128 it is apparent that

$$l(i_{bl}) \frac{di_{bl}}{dt} = v_{bl} = -\frac{\partial S}{\partial i_{bl}}$$

$$c(v_{bc}) \frac{dv_{bc}}{dt} = i_{bc} = \frac{\partial S}{\partial v_{bc}} \qquad (2.130)$$

These are the Brayton-Moser equations. The number of these first-order differential equations equals the number of inductors and capacitors in the network. The equations are sufficient to define the problem provided that the inductor currents and capacitor voltages form a complete set. Unfortunately this is frequently not true. The network can always be modified to make it true by adding series inductors and parallel capacitors. This is not unrealistic since such parasitic elements are always present. One difficulty with the method is the requirement to express S as a function of the currents in the inductors and the voltages across the capacitors so the operations in Equations 2.130 have meaning. If these form a complete set of network variables they define a variable in each branch uniquely and the calculation of S by Equation 2.129 is always possible theoretically. The actual calculation, however, is frequently awkward.

The preceding sections have discussed methods of networks analysis due to Hamilton, Lagrange, Kirchhoff, Bryant-Stern, and Brayton-Moser. Before proceeding with further discussion it is probably wise to illustrate and compare these methods by applying them to typical networks.

Figure 2.15 shows a conservative network in which it is assumed that the inductor l_1 is nonlinear. Applying the Hamilton method it is first necessary to form the Lagrangian

$$L = \frac{c_1 \dot{\phi}_1^2}{2} - \int_0^{\phi_1} i_{l_1}\, d\phi_1 + \frac{c_2 \dot{\phi}_2^2}{2} - \frac{(\phi_1 - \phi_2)^2}{2l_2} \qquad (2.131)$$

from which the generalized momenta are found to be

$$p_1 = \frac{\partial L}{\partial \dot{\phi}_1} = c_1 \dot{\phi}_1 = q_1$$

$$p_2 = \frac{\partial L}{\partial \dot{\phi}_2} = c_2 \dot{\phi}_2 = q_2 \qquad (2.132)$$

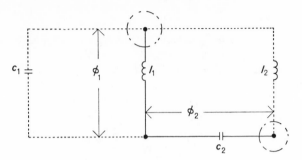

Figure 2.15 A conservative network.

The Hamiltonian is found by the relation

$$H = \sum q_s \dot{\phi}_s - L$$
$$= \frac{q_1^2}{2c_1} + \int_0^{\phi_1} i_{l_1}\, d\phi_1 + \frac{q_2^2}{2c_2} + \frac{(\phi_1 - \phi_2)^2}{2l_2} \tag{2.133}$$

and the canonical equations give

$$\dot{\phi}_1 = \frac{\partial H}{\partial q_1} = \frac{q_1}{c_1}$$

$$\dot{q}_1 = -\frac{\partial H}{\partial \phi_1} = -i_{l_1}(\phi_1) - \frac{\phi_1 - \phi_2}{l_2}$$

$$\dot{\phi}_2 = \frac{\partial H}{\partial q_2} = \frac{q_2}{c_2} \tag{2.134}$$

$$\dot{q}_2 = -\frac{\partial H}{\partial \phi_2} = \frac{\phi_1 - \phi_2}{l_2}$$

The first and third equations merely relate variables through branch relations. The second and fourth equations are current sums for the two independent cut-sets which are shown in the figure.

Applying the Lagrange method to the same network and using ϕ_1 and ϕ_2 as the generalized coordinates give the two equations

$$c_1 \ddot{\phi}_1 + i_{l_1}(\phi_1) + \frac{\phi_1 - \phi_2}{l_2} = 0$$

$$c_2 \ddot{\phi}_2 - \frac{\phi_1 - \phi_2}{l_2} = 0 \tag{2.135}$$

These are current sums for the same cut-sets. They are second-order equations which are the direct result of eliminating q_1 and q_2 from the Hamilton equations.

Although the Hamilton method yields first-order equations its scope is limited because it applies only to conservative systems and it cannot account for the presence of nonenergic devices such as transformers. Figure 2.16 shows a network with a transformer and a load resistor which is assumed to be nonlinear. When applying the Lagrangian method to such a network it is necessary to select a maximal tree and the complete set of network variables such that a parallel circuit with a flux as the variable appears on one side of the transformer and a series circuit with a charge as the variable appears on the other side. This provides proper variables for the Lagrangian function for the transformer. Using the variables shown in Figure 2.16 the Lagrangian for the network is

$$L = \frac{c_1 \dot{\phi}_1^2}{2} + n\dot{\phi}_1 q_2 + \frac{l\dot{q}_3^2}{2} - \frac{(q_2 - q_3)^2}{2c_2} \tag{2.136}$$

and the hybrid dissipation function is

$$G_{iv} = \int_0^{v_1} i_{r_1} \, dv_1 + \int_0^{i_2} v_{r_2} \, di_2$$

$$= \frac{v_1^2}{2r_1} + \int_0^{i_2} v_{r_2} \, di_2 \tag{2.137}$$

Using ϕ_1, q_2, and q_3 as the three generalized coordinates the corresponding Lagrange equations are

$$c_1 \ddot{\phi}_1 + n\dot{q}_2 = i_1 - \frac{v_1}{r_1}$$

$$-n\dot{\phi}_1 + \frac{q_2 - q_3}{c_2} = -v_{r_2}(i_2) \tag{2.138}$$

$$l\ddot{q}_3 - \frac{q_2 - q_3}{c_2} = 0$$

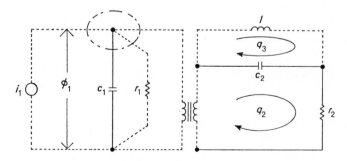

Figure 2.16 A network tractable by the Lagrange method.

The first equation is a current sum for the indicated cut-set and the last two equations are voltage sums around the basic meshes. It is interesting to observe that the inclusion of the term $n\dot{\phi}_1 q_2$ in the Lagrangian introduces the effect of the transformer properly into the network equations.

To demonstrate the comparative merit of the various methods of arriving at the network equations the network shown in Figure 2.17(a) is analyzed four different ways. Part (b) of the figure shows a maximal tree and three basic meshes. The complete set of variables is q_1, q_3, and q_5. Forming a sum

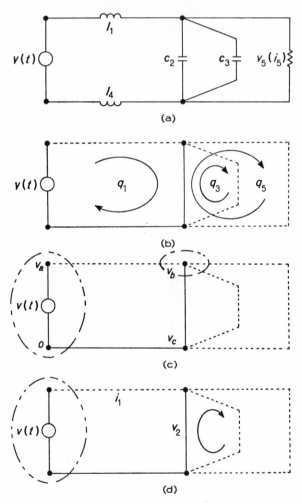

Figure 2.17   Different methods of analysis. (a) The network, (b) mesh analysis, (c) node analysis, and (d) hybrid analysis.

of voltages around each of the basic meshes provides the equations

$$l_1\ddot{q}_1 + \frac{q_1 - q_3 - q_5}{c_2} + l_4\ddot{q}_1 - v(t) = 0$$

$$\frac{q_3}{c_3} - \frac{q_1 - q_3 - q_5}{c_2} = 0 \tag{2.139}$$

$$v_5(q_5) - \frac{q_1 - q_3 - q_5}{c_2} = 0$$

Mesh analysis provides three equations in as many unknowns. The use of mesh charges as the variables eliminates integrals from the equations.

Part (c) shows an arbitrary assignment of potentials to the nodes of the network. Enough are assigned to determine the voltage across each tree branch. Then a flux ϕ_k is defined for each v_k. These fluxes are not necessarily related to the fluxes in the inductors but are directly related to the node potentials by $\dot{\phi}_k = v_k$. These fluxes are then used to express the branch voltages and the branch relations are then used to find the branch currents. The network equations are then generated by relating the source potential to the node potential and by forming appropriate cut-sets. Using the cut-sets shown in Figure 2.17(c) the equations are

$$\dot{\phi}_a - v(t) = 0$$

$$\frac{\phi_a - \phi_b}{l_1} - \frac{\phi_c}{l_4} = 0 \tag{2.140}$$

$$\frac{\phi_b - \phi_a}{l_1} + (c_2 + c_3)(\ddot{\phi}_b - \ddot{\phi}_c) + i_5 = 0$$

In this particular network node analysis also generates three equations. It is probably true, however, that with the majority of networks of typical or commonly occurring configuration mesh analysis yields fewer equations.

Part (d) of Figure 2.17 shows the selection of variables for the Bryant-Stern method. The dynamic set of network variables consists of i_1 and v_2. The inductive cut-set gives $i_4 = i_1$ and the capacitive mesh gives $v_3 = v_2$. Forming a voltage sum around the basic mesh corresponding to the inductive chord gives

$$\dot{\phi}_1 + v_2 + \dot{\phi}_4 - v(t) = 0 \tag{2.141}$$

Letting $\phi = \phi_1 + \phi_4 = (l_1 + l_4)i_1$ yields the relation

$$i_1 = \frac{\phi}{l_1 + l_4} \tag{2.142}$$

Forming a current sum through the cut-set corresponding to the capacitive tree branch gives

$$i_1 - \dot{q}_2 - \dot{q}_3 - i_5 = 0 \qquad (2.143)$$

Letting $q = q_2 + q_3 = (c_2 + c_3)v_2$ yields the relation

$$v_2 = \frac{q}{c_2 + c_3} \qquad (2.144)$$

Equation 2.143 involves i_5 instead of $v_5(i_5)$ so for simplicity the function $f(q)$ will be defined as

$$i_5 = i_5(v_5) = i_5(v_2) = f(q) \qquad (2.145)$$

Substituting for ϕ_1, ϕ_4, q_2, q_3, i_1, i_5, and v_2 in Equations 2.141 and 2.143 gives the final result

$$\dot{\phi} = -\frac{q}{c_2 + c_3} + v(t)$$

$$\dot{q} = +\frac{\phi}{l_1 + l_4} - f(q) \qquad (2.146)$$

These equations should be compared with Equations 2.139 and 2.140 to observe the advantage in hybrid analysis by the Bryant-Stern method.

Part (d) of Figure 2.17 can also be used in the Brayton-Moser method. To compute the mixed potential function it is necessary to know i_5 explicitly so a function $g(v_2)$ is defined by the relation $g(v_2) = i_5$. Then the application of Equation 2.129 permits the mixed potential function to be expressed as

$$S = -(i_2 + i_3)v_2 + \int_0^{g(v_2)} v_2 \, dg(v_2) - \int_0^{i_1} v(t) \, di_1 \qquad (2.147)$$

Applying Kirchhoff's current law and integrating the first integral by parts give the alternate expression

$$S = [i_1 - g(v_2)]v_2 + v_2 g(v_2) - \int_0^{v_2} g(v_2) \, dv_2 - \int_0^{i_1} v(t) \, di_1$$

$$= i_1 v_2 - \int_0^{v_2} g(v_2) \, dv_2 - \int_0^{i_1} v(t) \, di_1 \qquad (2.148)$$

and applying the Brayton-Moser formulas given by Equations 2.130 provides the final result

$$(l_1 + l_4)\frac{di_1}{dt} = -v_2 + v(t)$$

$$(c_2 + c_3)\frac{dv_2}{dt} = i_1 - g(v_2) \qquad (2.149)$$

The sums of the inductances and capacitances must be used because they are connected in series and parallel respectively to the resistive portion of the network. These equations are the same as the Bryant-Stern equations. This is not always true but in this simple example the dynamic set of network variables equals the complete set. Normally the Bryant-Stern method produces fewer and simpler equations.

HOMOGENEOUS MULTIPORT ELEMENTS

A multiport element is one which has numerous terminal pairs or nodes to which external connections can be made. It is homogeneous if it contains network elements of only one kind. This section describes some properties of such elements. Theorems 2.7 through 2.9 regarding content and cocontent are the governing considerations for nonlinear resistive networks. The proof of Theorem 2.9 is not limited to a single port so the theorem applies equally well to multiport elements. For the internal content and cocontent to be stationary it is necessary that the external sources at all ports remain constant.

There are some additional theorems about homogeneous multiport elements which are concerned primarily with inductive and capacitive networks. These are more easily established by introducing Kirchhoff's laws in a different form.

Postulate 2.3   *In a given time interval the total flux which changes around a mesh is zero.*

Postulate 2.4   *In a given time interval the total charge which changes at a node is zero.*

Using these postulates plus the previous ones it is possible to extend Tellegen's theorem to purely inductive or capacitive networks that are complete.

Theorem 2.10   *In a complete Kirchhoffian network the sum of the products of the current and the flux in each branch is zero.*

Assuming that the network satisfies Kirchhoff's current law then the sum can be expressed as

$$\sum_b i_b \phi_b = \frac{1}{2} \sum_j \sum_k (\phi_j - \phi_k) i_{jk}$$

$$= \sum_j \phi_j \sum_k i_{jk}$$

$$= 0 \qquad\qquad (2.150)$$

The theorem may also be established by using mesh analysis, utilizing Postulate 2.3, and assuming that all flux is initially zero. The proof of the following dual theorem is left as an exercise.

Theorem 2.11 *In a complete Kirchhoffian network the sum of the products of the charge and the voltage in each branch is zero.*

The proofs of the following theorems are identical to that of Theorem 2.7 except for changes in variables and the use of Postulates 2.3 and 2.4 so they are stated here without proof.

Theorem 2.12 *If a complete inductive network satisfies Kirchhoff's laws then magnetic energy and coenergy are conserved.*

Theorem 2.13 *If a complete capacitive network satisfies Kirchhoff's laws then electric energy and coenergy are conserved.*

Since complete homogeneous networks have the above properties of conserving various complementary quantities it is to be expected that these quantities exhibit stationary values within such networks under the stipulated conditions. The proofs of the following theorems are similar to that of Theorem 2.9 and are not repeated.

Theorem 2.14 *In any inductive network the magnetic coenergy is stationary against current variations and the magnetic energy is stationary against voltage variations.*

Theorem 2.15 *In any capacitive network the electric energy is stationary against current variations and the electric coenergy is stationary against voltage variations.*

The preceding theorems about the complementary functions assume that the functions are uniquely defined by the branch variables and that they therefore are suitable state functions. It is perhaps advisable to investigate the conditions under which this is true. It is assumed that x_i and x_j are the independent branch variables at ports i and j respectively. The dependent variables are y_i and y_j. The criterion for y_i and y_j to produce suitable state functions is that when the conditions at each port pass from state A to state B the evaluation of the state functions depends only on the endpoints of the intervals of x_i and x_j. Referring to Figure 2.18(a) this is equivalent to the requirement that

$$\sum \int_{APB} y \, dx = \sum \int_{AQB} y \, dx \tag{2.151}$$

where the sum is formed over both ports. This is equivalent to

$$\sum \int_{APB} y \, dx + \sum \int_{BQA} y \, dx = 0 \tag{2.152}$$

Therefore the criterion can be stated as the requirement that the line integral around a closed curve is zero. Due to cancellation of the contributions along

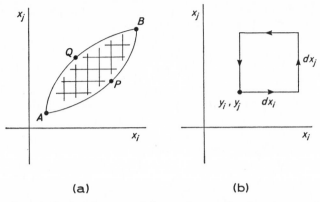

Figure 2.18 Development of the requirement for the existence of state functions and reciprocity.

opposing internal paths as shown in Figure 2.18(a) a necessary and sufficient requirement is that the line integral is zero around every small square. Figure 2.18(b) shows such a square. The sum of the integrals at both ports is given by

$$\sum \oint y \, dx = y_i \, dx_i + \left(y_j + \frac{\partial y_j}{\partial x_i} \, dx_i \right) dx_j$$

$$- \left(y_i + \frac{\partial y_i}{\partial x_j} \, dx_j \right) dx_i - y_j \, dx_j$$

$$= \left(\frac{\partial y_j}{\partial x_i} - \frac{\partial y_i}{\partial x_j} \right) dx_i \, dx_j \qquad (2.153)$$

and the condition for a null integral is that

$$\frac{\partial y_j}{\partial x_i} = \frac{\partial y_i}{\partial x_j} \qquad (2.154)$$

This necessary and sufficient condition for the existence of state functions defines a property of multiport elements known as *reciprocity*. This property is intrinsic with passive linear systems containing ordinary elements. The gyrator, however, is not reciprocal. Nonlinear networks containing elements which display hysteresis or unilateral characteristics are also not reciprocal. In special cases it is possible for a network to be reciprocal between only a few of its several ports. Such cases must be treated with caution.

RECTANGLE DIAGRAMS

Certainly the most common method of representing networks of physical devices is to use a mathematical graph with an appropriate symbol in each branch to designate the type of device which is present in that branch. Cherry and others have proposed a completely different representation which provides an intuitive approach to network problems and which has some value in determining equilibrium conditions and other properties of homogeneous networks. The method applies directly to passive planar networks but it can be extended to treat more general situations. In this alternate representation the network appears as a rectangle and the network elements appear as smaller rectangles lying within the larger one. The fact that they exactly fill the larger rectangle depends on two properties of Kirchhoffian networks. First, if the source or sources are excluded then any cut-set through the passive portion of the network gives a constant value of current which is that supplied by the sources. Second, the sum of the voltages along any paths between two nodes gives a constant value of voltage which is the voltage between the nodes.

Figure 2.19(a) shows a simple network. The corresponding rectangle diagram is shown in part (b). Voltage is depicted vertically and current is shown horizontally. The line from P to Q represents a cut-set through resistors 1 and 2. The fact that the diagram has the same width across resistor 3 demonstrates Kirchhoff's current law. The line from R to S has a length corresponding to the voltage v and this demonstrates Kirchhoff's voltage law.

The slopes of the diagonals across each of the small rectangles are the resistances of each of the resistors and the slope of the diagonal across the large rectangle is the effective resistance of the network. If the resistors are linear then a change of voltage merely changes the size of the diagram but the relationships remain unchanged. If the resistors are nonlinear a change in voltage will usually cause the relative sizes of the rectangles to vary. For a resistive network the diagram demonstrates conservation of power since the dissipation of each resistor equals the area of its rectangle and the power supplied by the source equals the area of the large rectangle.

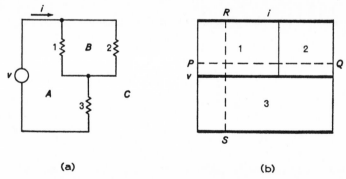

(a)   (b)

Figure 2.19   A simple network and its rectangle diagram.

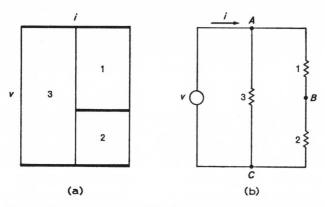

Figure 2.20 Development of the dual network. (a) The rotated rectangle diagram and (b) the corresponding network.

The dual of a network is constructed by assigning a node to each simple mesh and joining nodes whenever the meshes have a branch in common. Under this transformation mesh currents become node voltages and branch voltages become branch currents. This exchange of voltages and currents can be accomplished easily with the rectangle diagram. As shown in Figure 2.20(a) the diagram is rotated 90° and the voltages and currents are exchanged. The corresponding network is shown in Figure 2.20(b). The lettering of the nodes matches that of the meshes of Figure 2.19(a).

Part (a) of Figure 2.21 shows a Kuratowski hexoid. It is demonstrated in the first chapter that this is a nonplanar graph. This means that it cannot be drawn on a plane surface without intersecting branches. Figure 2.21(b)

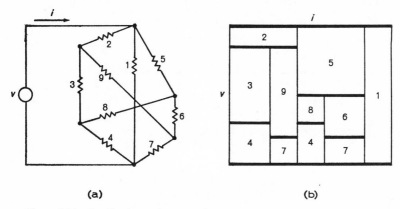

Figure 2.21 Application of rectangle diagram to nonplanar graphs. (a) Kuratowski hexoid and (b) the corresponding rectangle diagram.

shows a rectangle diagram for this network. The requirement for dividing some of the smaller rectangles into several sections removes almost all practical value from this construction.

The rectangle diagram can represent networks containing sources. If, for example, the sources within the portion of the network which is represented by the rectangle diagram act so as to reinforce the external generator then the voltages and currents have senses opposed to those of the resistors. The resultant diagram requires the use of overlapping rectangles but it follows the same rules and it is not difficult to construct.

Cherry has used the rectangle diagram to show how it is possible to use nonlinear resistors in a four-terminal bridge or lattice network in such a manner that the network presents a linear input resistance which always equals its load resistance for any input voltage. Figure 2.22(a) shows the network and part (b) shows the rectangle diagram. Under the required conditions the slope of the diagonal of the large rectangle must always be equal to the load resistance r. Geometric relationships in the rectangle diagram show that it must always be true that

$$\frac{v_2}{i_1} = \frac{v_1}{i_2} = r \qquad (2.155)$$

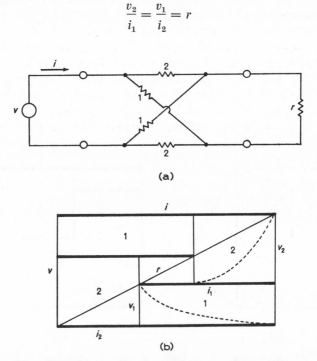

Figure 2.22 Linear iterative resistance using nonlinear resistors. (a) The lattice network and (b) the rectangle diagram.

These equations may be written in the form

$$\left(\frac{v_2(i_2)}{i_2}\right) = r^2\left(\frac{i_1(v_1)}{v_1}\right) \tag{2.156}$$

Assuming that r is constant this equation states that the requirement for an input resistance to be equal to the load resistance is equivalent to the requirement that the two sets of resistors have dual characteristics plus the incorporation of the square of the load resistance as a scale factor. Numerous networks of this type can be connected in cascade without changing the load placed on the source. The recurring resistance r is called the *iterative resistance* and such a connection provides unlimited linear attenuation using nonlinear elements.

All of the previous examples use resistive networks. The rectangle diagram can be used to represent any homogeneous network by changing variables. Flux and current are used in inductive networks while charge and voltage are used in capacitive networks. Since the rectangle diagram describes the state of the network under stationary conditions using only two variables it obviously does not apply to networks containing more than one type of element.

REFERENCES

1. T. E. Stern, *Theory of Nonlinear Networks and Systems*, Addison-Wesley, Reading, 1965.
2. C. Lanczos, *The Variational Principles of Mechanics*, University of Toronto Press, Toronto, 1949.
3. B. R. Gossick, *Hamilton's Principle and Physical Systems*, Academic Press, New York, 1967.
4. S. Seshu and M. B. Reed, *Linear Graphs and Electrical Networks*, Addison-Wesley, Reading, 1961.
5. S. Duinker, Search for a Complete Set of Basic Elements for the Synthesis of Nonlinear Electrical Systems, *Recent Developments in Network Theory*, Macmillan, New York, 1963.
6. C. Cherry, Some General Theorems for Nonlinear Systems Possessing Reactance, *Philosophical Magazine*, Volume 42, 1951, pp. 1161–1177.
7. W. Millar, Some General Theorems for Nonlinear Systems Possessing Resistance, *Philosophical Magazine*, Volume 42, 1951, pp. 1150–1160.
8. S. Duinker, Generalization to Nonlinear Networks of a Theorem due to Heaviside, *Philips Research Reports*, Volume 14, 1959, pp. 421–426.
9. W. H. Ingram and C. M. Cramlet, On the Foundations of Electrical Network Theory, *Journal of Mathematics and Physics*, Volume 23, 1944, pp. 134–155.
10. J. L. Synge, The Fundamental Theorem of Electrical Networks, *Quarterly of Applied Mathematics*, Volume 9, 1951, pp. 113–127.
11. C. Saltzer, The Second Fundamental Theorem of Electrical Networks, *Quarterly of Applied Mathematics*, Volume 11, 1953, pp. 119–123.

12. G. Kirchhoff, On the Solution of the Equations Obtained from the Investigation of the Linear Distribution of Galvanic Currents, *IRE Transactions on Circuit Theory*, Volume CT-5, 1958, pp. 4–7, (translation by J. B. O'Toole).

13. R. K. Brayton and J. K. Moser, A Theory of Nonlinear Networks, *Quarterly of Applied Mathematics*, Volume 22, 1964, pp. 1–33, 81–104.

14. P. R. Bryant, The Order of Complexity of Electrical Networks, Monograph 335 E, *Institution of Electrical Engineers*, 1959, pp. 174–188.

15. T. E. Stern, On the Equations of Nonlinear Networks, *IEEE Transactions on Circuit Theory*, Volume CT-13, 1966, pp. 74–81.

16. B. D. H. Tellegen, A General Network Theorem with Applications, *Philips Research Reports*, Volume 7, 1952, pp. 259–269.

17. C. Cherry, Classes of Four-Pole Networks Having Nonlinear Transfer Characteristics but Linear Iterative Impedance, *Proceedings of the IEE*, Volume 107B, 1960, pp. 26–30.

3

POLYNOMIALS AND SINUSOIDS

A nonlinear device is usually defined as one wherein the characteristic curve relating the output to the input is not a straight line. Therefore if several input signals are combined linearly the output is not necessarily the linear combination of the respective outputs and superposition cannot be used.

Another aspect of its behavior is rather thoroughly described in this chapter. If the input signal is a pure sinusoidal vibration then the output always contains vibrations whose frequencies are harmonically related to the frequency of the input signal. This behavior is inherent in all nonlinear processes.

The first part of this chapter gives selected numerical and graphical methods for representing the characteristic curve as a polynomial. The next part links this representation with the generation of harmonic frequencies. In particular, the mathematical device known as Pascal's triangle is introduced as a mnemonic aid.

The next section gives methods for decomposing a wave into its various components. There is a rather complete description of Fourier series. A subsequent section treating Fejér series shows that Fourier series is neither optimum nor unique. The chapter closes with the development of a method which uses Chebyshev polynomials to predict harmonic amplitudes directly.

POLYNOMIAL EVALUATION

Methods for converting information to polynomial form are usually based on a reversal of the procedures for evaluating a polynomial. Numerical methods for achieving this are well known, hence only graphical techniques will be given here. These methods, which are due to Lill and Segner, are

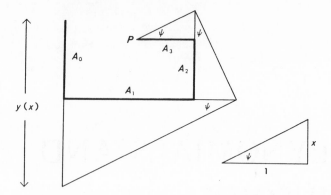

Figure 3.1 The Lill method for evaluating a polynomial.

described in great detail in the book by Willers which is listed as a reference. It is assumed that the polynomial is expressed in the general form

$$y(x) = A_0 + A_1 x + A_2 x^2 + \cdots + A_n x^n \qquad (3.1)$$

Figure 3.1 shows how the Lill method is used to evaluate a cubic polynomial. First, the coefficients are measured along a sequence of lines each rotated 90° with respect to the previous one. The order is from highest to lowest subscript. If a coefficient is zero a rotation of 180° is used. Next, the auxiliary

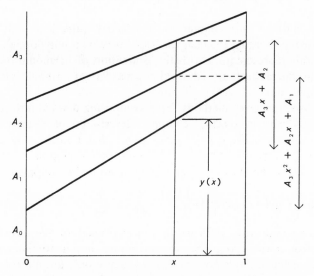

Figure 3.2 The Segner method for evaluating a polynomial.

triangle is drawn to find ψ. The second sequence of lines is now drawn also starting from the point P. The successive operations of multiplication and addition finally synthesize the complete polynomial. The value of $y(x)$ can be found by measuring the indicated line segment.

The Segner method for evaluating a polynomial is shown in Figure 3.2. Multiplication in this method is also performed by the use of similar triangles. Along the vertical axis on the left in Figure 3.2 the coefficients A_i are measured as a sequence of separate line segments as shown. They are not measured from the origin. The values of the various line segments including $y(x)$ are shown. For pictorial clarity all of the coefficients are positive and $x < 1$. The same relationships are maintained in the more general case.

INTERPOLATION

This section describes procedures wherein a polynomial is constrained to fit a given function at a few prescribed points. Since the results are frequently used to evaluate the function at undefined intermediate points the general theory is listed in mathematics literature as interpolation. The theoretical justification which underlies all of this theory is the remarkable and important approximation theorem given in 1885 by Weierstrass.

Theorem 3.1 *If $F(x)$ is real and continuous in a closed interval then, given $\epsilon > 0$, there exists a polynomial $y(x)$ such that within the interval $|F(x) - y(x)| < \epsilon$.*

A proof of this theorem is outlined later in this chapter following the development of Fejér series. The theorem is applied to nonlinear analysis by considering $F(x)$ to be the continuous function truly describing the nonlinear device. If the basic information is in tabular form it is assumed at this point that this information has been smoothed to remove irrelevant fluctuations due to experimental error.

The polynomial can be constrained to pass *exactly* through the arbitrary points shown in Figure 3.3 by writing it in the form

$$y(x) = a_0 + a_1(x - x_0) + a_2(x - x_0)(x - x_1)$$
$$+ \cdots + a_n(x - x_0)(x - x_1) \cdots (x - x_{n-1}) \quad (3.2)$$

As x takes successive values from x_0 to x_{n-1} there is generated the sequence of equations

$$y_0 = a_0$$
$$y_1 = a_0 + a_1(x_1 - x_0) \quad\quad\quad\quad\quad (3.3)$$
$$y_2 = a_0 + a_1(x_2 - x_0) + a_2(x_2 - x_0)(x_2 - x_1)$$

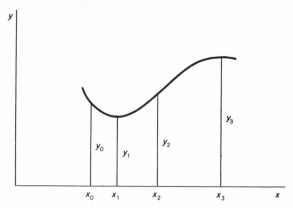

Figure 3.3   Fitting a polynomial to a set of points.

The constants a_i may be evaluated in order. After the a_i have been determined, the A_i are found by equating coefficients of like powers in Equations 3.1 and 3.2. For a cubic polynomial

$$A_0 = -[a_3(x_2 x_1 x_0) - a_2(x_1 x_0) + a_1(x_0) - a_0]$$

$$A_1 = +[a_3(x_2 x_1 + x_1 x_0 + x_0 x_2) - a_2(x_1 + x_0) + a_1] \qquad (3.4)$$

$$A_2 = -[a_3(x_2 + x_1 + x_0) - a_2]$$

$$A_3 = +[a_3]$$

The coefficients for a quadratic polynomial are found by letting $a_3 = 0$. Any desired higher order relations are derived by repeating the above process with the required higher order terms. The results have the same form as Equations 3.4 but increase in complexity rapidly with increasing order. For all orders $A_n = a_n$ so the process must be continued until the highest order coefficient is essentially zero. Extensions to higher order are easily achieved by selecting additional values from the original graphical or tabular information. This is possible since there has been no restriction on the positions of the points or the total number of points.

It is easy to perform interpolation graphically. Figure 3.4 shows how the Lill method may be used. First, the values of selected y_i are measured along a straight line from the point Q. Next, triangles are constructed to find all required ψ_{ij}. These correspond to the various differences in x which appear in Equations 3.3 and determine the orientations of the construction lines. The lines are drawn in the order indicated by the encircled numbers. As the construction proceeds, the sequence given in Equations 3.3 is solved for the a_i.

The use of the Segner technique to solve the interpolation problem is shown in Figure 3.5. It solves the same set of equations as the Lill method and

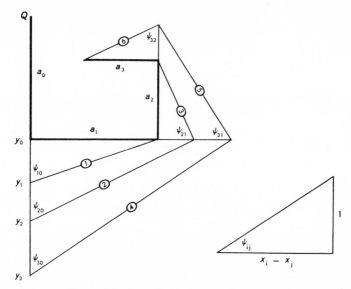

Figure 3.4 Interpolation by the Lill method.

it requires the same number of construction lines. Their order is again indicated by the encircled numbers.

If the interpolation is to be performed by numerical rather than graphical methods it is much easier to use equal increments in the independent variable. If the distance between any two adjacent values of x is called h then Equations 3.3 become

$$y_0 = a_0$$
$$y_1 = a_0 + a_1 h \qquad\qquad (3.5)$$
$$y_2 = a_0 + 2a_1 h + 2a_2 h^2$$

These may be solved sequentially for the interpolation coefficients to give

$$a_0 = y_0$$
$$a_1 = \frac{y_1 - y_0}{h} = \frac{\Delta y_0}{h}$$
$$a_2 = \frac{\Delta y_1 - \Delta y_0}{2h^2} = \frac{\Delta^2 y_0}{2h^2} \qquad\qquad (3.6)$$
$$a_n = \frac{\Delta^n y_0}{n!\, h^n}$$

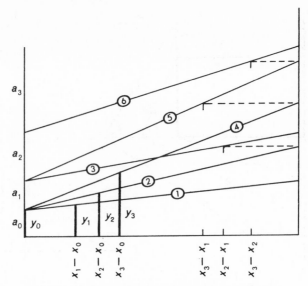

Figure 3.5   Interpolation by the Segner method.

A careful analysis of the above procedure shows that the higher order differences involve increasingly large numbers of the y_i values and can be conveniently calculated as shown below. Each new entry in the table is the difference between adjacent entries to the left.

$$
\begin{array}{llll}
x_0 & y_0 & & \\
 & & \Delta y_0 & \\
x_1 & y_1 & & \Delta^2 y_0 \\
 & & \Delta y_1 & & \Delta^3 y_0 \\
x_2 & y_2 & & \Delta^2 y_1 \\
 & & \Delta y_2 & \\
x_3 & y_3 & &
\end{array}
$$

To construct a polynomial which fits the prescribed points it is sufficient to use the entries at the top of each column to calculate the interpolation coefficients by the use of Equations 3.6. The coefficients in the required polynomial are then found using Equations 3.4.

The above procedure will be illustrated with an example. In the table below, the y_i correspond to the Briggsian logarithms of the x_i. This function was selected because it is smooth and continuous yet it bears no direct relationship to a polynomial.

x	y	Δy	$\Delta^2 y$	$\Delta^3 y$
20	1.3010			
		0.1761		
30	1.4771		−0.0511	
		0.1250		0.0240
40	1.6021		−0.0271	
		0.0979		0.0094
50	1.6990		−0.0177	
		0.0792		0.0054
60	1.7782		−0.0123	
		0.0669		0.0034
70	1.8451		−0.0089	
		0.0580		0.0020
80	1.9031		−0.0069	
		0.0511		
90	1.9542			

It is assumed that it is desired to fit a section of the table to a high degree of accuracy with a cubic polynomial so differences up to the third order are indicated. These are rather large down to the last two entries so the permitted range of x will be from 50 to 90. Inserting the underlined differences from the table into Equations 3.6 gives

$$a_0 = 1.6990$$

$$a_1 = \frac{\Delta y_0}{h} = 0.00792$$

$$a_2 = \frac{\Delta^2 y_0}{2h^2} = -0.000062 \qquad (3.7)$$

$$a_3 = \frac{\Delta^3 y_0}{6h^3} = 0.0000006$$

Substitution of these values and appropriate x_i in Equations 3.4 gives the coefficients of the desired cubic polynomial

$$y(x) = 0.9910 + 0.02116x + 0.000170x^2 + 0.0000006x^3 \qquad (3.8)$$

After achieving a theoretically perfect fit at a set of prescribed points it is interesting to learn how well the polynomial approximates the original function at intermediate points. A comparison of these calculations with the values of the function appearing in the original logarithm table is shown below. A small, typical oscillation of the cubic function is observed in the last two rows.

x	Tabular values	Calculated values
55	1.7404	1.7404
65	1.8129	1.8129
75	1.8751	1.8748
85	1.9294	1.9298

There are errors almost always in the original tabular information. They are frequently due to experimental limitations but may also be caused by standardizing the entries in the table to a fixed number of decimal places. If there is an error in only one of the entries then this error propagates as follows:

y	Δy	$\Delta^2 y$	$\Delta^3 y$
—	—		
	—		ϵ
—		ϵ	
	ϵ		-3ϵ
ϵ		-2ϵ	
	$-\epsilon$		3ϵ
—		ϵ	
	—		$-\epsilon$
—	—		

Since the differences tend to decrease and the propagated errors tend to increase, such an error can usually be detected by forming differences until the values in a column oscillate about a mean value. The location and amplitude of the oscillation indicate the position and size of ϵ. This type of error is the best situation that can arise.

The worst case occurs when the errors in the original information alternate in sign down a column. In this case the propagated errors appear as follows:

y	Δy	$\Delta^2 y$	$\Delta^3 y$
ϵ		-4ϵ	
	-2ϵ		8ϵ
$-\epsilon$		4ϵ	
	2ϵ		-8ϵ
ϵ		-4ϵ	
	-2ϵ		8ϵ
$-\epsilon$		4ϵ	
	2ϵ		-8ϵ
ϵ		-4ϵ	

They again oscillate in sign down a column. If r is the order of the difference then the error in that column lies within $\pm 2^r \epsilon$. This is an upper bound on the error.

Usually the error will lie somewhere between the above values but it is clear that there is always a limit to the extent of the process. Beyond this point the propagated errors mask any significant trends.

The interpolation method which has been presented above uses one form of the Gregory-Newton formula. This was the first to be given (1670) and uses the most direct approach. Others are easily derived. There are numerous identities involving entries in the table. Differences may also be formed in a backward direction from the bottom of the table to the top. Average values may be used. These techniques lead to other formulas bearing such names as Gauss, Bessel, Everett, Steffensen, and Stirling although there is some confusion regarding true authorship. There is no single formula which is superior to all others in all cases. For the analysis of the characteristics of actual nonlinear devices the direct method outlined above is usually adequate.

LEAST-SQUARES APPROXIMATIONS

The preceding methods fit a polynomial precisely to a given set of points. The required number of points equals the number of coefficients in the polynomial. If there are additional points available this excess information can be used to smooth any apparently irrelevant fluctuations in the basic information or it can be ignored.

This arbitrary smoothing or disregard of available information is avoided by the use of a least-squares technique. In this method it is assumed that the true characteristic $F(x)$ is expressed either in graphical or tabular form and is to be approximated by a function of the form

$$y(x) = \sum_{k=0}^{n} c_k \phi_k(x) \qquad (3.9)$$

where each ϕ_k is a polynomial of order k. The criterion for determining both the forms of the polynomials and the magnitudes of the coefficients is based on minimizing the aggregate of the square of the error between $F(x)$ and $y(x)$. This requirement is made more useful by providing a weighting function which assigns varying importance to the error depending on its position in the interval. It will be observed later that this weighting function also provides better convergence of the expressions to be developed.

The least-squares criterion with weighting function $p(x)$ demands that

$$\int_a^b p(x)[F(x) - y(x)]^2 \, dx$$

be minimized by a proper choice for $y(x)$. Assuming at this point that $y(x)$ is a function of the expansion coefficients then the extremum for the above integral is obtained by a simultaneous solution of the $(n + 1)$ relations of the type

$$\frac{\partial}{\partial c_r} \int_a^b p(F - y)^2 \, dx = 0 \tag{3.10}$$

By substituting the sum for y and performing the differentiation these are converted into equations such as

$$\sum_{k=0}^n c_k \int_a^b p\phi_r\phi_k \, dx = \int_a^b pF\phi_r \, dx \tag{3.11}$$

If a set of polynomials and the weighting function are specified then these may be solved simultaneously for the expansion coefficients.

The amount of effort required to complete the above calculation is reduced very much by selecting polynomials with a special property. If the functions satisfy the requirement that

$$\int_a^b p\phi_r\phi_k \, dx = 0 \qquad r \neq k \tag{3.12}$$

then the array of integrals in the sums in Equation 3.11 reduces to a single integral in each case and the expansion coefficients can be found by the much simpler formula

$$c_r = \frac{\displaystyle\int_a^b pF\phi_r \, dx}{\displaystyle\int_a^b p\phi_r^{\,2} \, dx} \tag{3.13}$$

To evaluate these integrals properly many values of $F(x)$ must be used.

Functions which have the property described in Equation 3.12 are said to be *orthogonal*. The name arises from the similarity in form and method for expressing an arbitrary vector in terms of mutually perpendicular vectors and for expressing an arbitrary function in terms of these functions. Orthogonality prohibits cross coupling in the calculations of the components and each evaluation is independent of the others.

It was necessary to assume the existence of orthogonal polynomials to derive Equation 3.13. If it is astonishing to learn that such polynomials exist at all, it will be even more astounding to learn that numerous distinct sets of such functions have been discovered. To generate these sets it is expedient to formulate general requirements for all sets. To do this it has been found convenient to introduce a generating function which is defined by the relation

$$p\phi_r = \frac{d^r}{dx^r} u_r = u_r^{(r)} \tag{3.14}$$

Substituting this into Equation 3.12, assuming that ϕ_k has a degree less than r, and integrating successively by parts give the boundary value requirements

$$u_r(a) = u'_r(a) = u''_r(a) = \cdots = u_r^{(r-1)}(a) = 0$$
$$u_r(b) = u'_r(b) = u''_r(b) = \cdots = u_r^{(r-1)}(b) = 0$$

(3.15)

To assure that ϕ_r has the degree r it must be the highest degree polynomial which satisfies the relation

$$\phi_r^{(r+1)} = \left[\frac{u_r^{(r)}}{p}\right]^{(r+1)} = 0$$

(3.16)

This differential equation combined with the above boundary conditions yields the desired orthogonal polynomials. The exact form of these polynomials depends on the limits of integration which are chosen and the weighting function which is used.

In developing the requirement stipulated in Equation 3.15 it was assumed that ϕ_k has a degree less than r. Otherwise ϕ_k is arbitrary. Therefore $p\phi_r$ is orthogonal to all polynomials with degree less than r and it is true that

$$\int_a^b p\phi_r^2 \, dx = \alpha_r \int_a^b p\phi_r x^r \, dx$$

(3.17)

where α_r is the coefficient of x^r in ϕ_r. Integrating by parts r times with the assumed boundary conditions leads to the useful result

$$\int_a^b p\phi_r^2 \, dx = (-1)^r r! \, \alpha_r \int_a^b u_r \, dx$$

(3.18)

This formula provides a means for evaluating the denominator in Equation 3.13.

If the interval from -1 to $+1$ is selected and if it is desired to place equal emphasis on all error and hence use the weighting function $p = 1$ then the solution of Equations 3.15 and 3.16 gives the generating function

$$u_r = C_r(x^2 - 1)^r$$

(3.19)

The constant C_r is usually chosen to yield polynomials in a normal form. This selection gives a set of functions known as *Legendre polynomials*. They are generated by the equation

$$P_k(x) = \frac{1}{2^k k!} \frac{d^k}{dx^k} (x^2 - 1)^k$$

(3.20)

The first four Legendre polynomials are

$$P_0 = 1$$
$$P_1 = x$$
$$P_2 = \frac{3x^2 - 1}{2}$$
$$P_3 = \frac{5x^3 - 3x}{2}$$

(3.21)

When the polynomials are written in this form Equation 3.18 provides an evaluation of the integral

$$\int_{-1}^{1} P_r^2 \, dx = \frac{2}{2r + 1}$$

(3.22)

which leads to a formula for the expansion coefficients

$$c_k = \frac{2k + 1}{2} \int_{-1}^{1} F(x) P_k(x) \, dx$$

(3.23)

It should perhaps be repeated that the least-squares technique is used only when a large number of points is available. This provides the information necessary to evaluate the above integral accurately.

When an approximation is made using Legendre polynomials the general form is described by Equation 3.9, the polynomials are defined by Equation 3.20, and the coefficients are found usually by numerical integrations to evaluate Equation 3.23. Therefore the normal forms of the Legendre polynomials are useful only in the interval from -1 to $+1$. If the nonlinear function is defined in another interval it is necessary to provide a translation and change of scale by introducing a new variable. The approximation is then performed with the new variable and the results are transformed back to the original variable to give the final result.

The use of an interval from zero to infinity and an exponential weighting function in the above procedure yields a set of orthogonal functions known as *Laguerre polynomials*. A doubly infinite interval and a second-order exponential weighting function produce *Hermite polynomials*. Although these polynomials have interesting theoretical properties their practical application is restricted due to the infinite intervals.

ALGEBRAIC APPROXIMATIONS

Before concluding this discussion of polynomials it will be shown how algebraic forms may be used to approximate some special nonlinear characteristics. In Figure 3.6 the dashed lines indicate the behavior of the algebraic

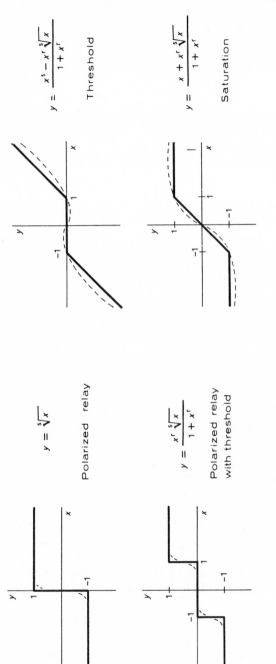

Figure 3.6 Approximations for some abrupt nonlinear characteristics.

expressions. It is assumed that $r = 2m$ and $s = 2m + 1$ so r is even and s is odd. The approximation improves as m increases. All expressions are normalized. These forms are of little direct value in the solution of practical problems but may offer some insight into the mechanisms of the processes. Other forms may be created by proper combinations of those shown.

HARMONIC GENERATION

The previous discussion has outlined various means for representing a nonlinear function as a polynomial. When the nonlinearity of a network element is represented in this form it is easy to describe what happens when sinusoidal vibrations are impressed on the device. The output is a distorted wave and it will be shown that the distortion indicates the presence of new vibrations with harmonically related frequencies. The analysis is facilitated by expressing the vibration in an exponential form and temporarily considering the double-amplitude input signals

$$\overline{\cos} \; \theta = 2 \cos \theta = e^{j\theta} + e^{-j\theta}$$
$$\overline{\sin} \; \theta = 2j \sin \theta = e^{j\theta} - e^{-j\theta}$$

(3.24)

where $\theta = \omega t$. Applying the cosine vibration to terms of successively higher order gives the sequence

$$\overline{\cos}^2 \theta = \qquad e^{j2\theta} + 2 + e^{-j2\theta} \qquad = \overline{\cos} \; 2\theta + 2$$

$$\overline{\cos}^3 \theta = \quad e^{j3\theta} + 3e^{j\theta} + 3e^{-j\theta} + e^{-j3\theta} \quad = \overline{\cos} \; 3\theta + 3 \overline{\cos} \; \theta \qquad (3.25)$$

$$\overline{\cos}^4 \theta = e^{j4\theta} + 4e^{j2\theta} + 6 + 4e^{-j2\theta} + e^{-j4\theta} = \overline{\cos} \; 4\theta + 4 \overline{\cos} \; 2\theta + 6$$

A similar sequence exists for the sine wave. Alternate terms are negative and odd orders contain sine vibrations.

It is observed that the coefficients form the same array of numbers as that in the mathematical device known as Pascal's triangle. The method of arriving at these numbers is slightly different but the results are identical so some properties of the triangle will be presented.

Two configurations of the array are shown below. Elements in the triangle on the right have the same numerical values but have associated with them the negative signs which occur when the input is a sine wave. The value of each element in the triangles is the sum of the values of the two neighboring elements in the row directly above if all algebraic signs are positive.

Since the outer elements are unity this method of construction yields elements whose numerical value also equals the number of possible paths from the apex of the triangle to the location of that element. If all of the paths are equally likely then the values of the elements along any row give the relative probabilities of arriving at random in the locations of those elements.

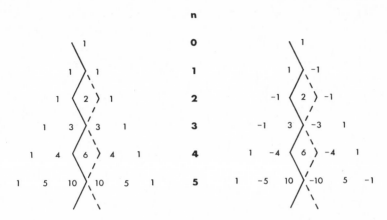

Figure 3.7   Use of Pascal's triangle to find harmonic amplitudes.

Since there are two alternatives at each point the total number of possibilities doubles with each increase in order and as a result the sum of the elements across each row is 2^n.

The elements in Pascal's triangle have another interesting and unexpected significance. The number of combinations of n things taken k at a time is calculated from the formula

$$_nC_k = \frac{n!}{k!\,(n-k)!} \tag{3.26}$$

From this relation there may immediately be derived the identity

$$_nC_k + {}_nC_{k+1} = {}_{n+1}C_{k+1} \tag{3.27}$$

This is precisely the rule by which the elements are calculated in Pascal's triangle if k is considered to be the order in which an element appears on a given row starting with zero order at the edge. This fact permits the harmonic amplitudes to be expressed in concise form. Such a relation has greater utility if it involves conventional sine and cosine functions rather than the double-amplitude vibrations which have been considered to this point. The incorporation of Equations 3.24 gives the final result

$$\cos^n \theta = \left(\frac{1}{2}\right)^n \sum_{k=0}^{n} {}_nC_k \cos (n - 2k)\theta$$

$$\sin^n \theta = \left(\frac{j}{2}\right)^n \sum_{k=0}^{n} (-1)^k {}_nC_k \cos (n - 2k)\theta \qquad \text{if } n \text{ is even} \tag{3.28}$$

$$\sin^n \theta = \frac{1}{j}\left(\frac{j}{2}\right)^n \sum_{k=0}^{n} (-1)^k {}_nC_k \sin (n - 2k)\theta \qquad \text{if } n \text{ is odd}$$

If the input is a cosine wave then the relative amplitudes in the output are found to the left of the solid line in the triangle on the left in Figure 3.7. The absolute harmonic amplitudes are found by dividing the elements in Pascal's triangle by 2^{n-1} with the exception of the central terms where $n = 2k$. These appear only once in the sums and must therefore be divided by 2^n. The frequencies start with a harmonic number equal to the order number and decrease successively by two. If the input is a sine wave then the algebraic signs must be used which are shown in the triangle on the right in Figure 3.7. Also, the output waves have a cosine phase if the wave is raised to an even power. As a consequence of these rules the following are true:

1. The fundamental frequency is missing in even powers but present in odd powers.

2. A constant is present in even powers but missing in odd powers.

3. The harmonic numbers are either all even or all odd in accordance with the power of the nonlinear term.

If the nonlinear process contains several orders then the output contains the sum of the vibrations which would be present individually. Superposition is permitted in this case because it is used in the construction of the nonlinear process by the algebraic addition of its components.

If the input wave has a phase angle Φ then $\theta = \omega t + \Phi$. When this is inserted in the previous formulas it is observed that the phase of the harmonic vibrations increases with the harmonic number in direct proportion.

If the input wave has many frequencies then it may be expressed as

$$x(t) = \sum_{k=-m}^{m} C_k e^{j\theta_k} \tag{3.29}$$

where m is the number of frequencies present. If the wave is raised to the power n it must be expressed by the redoubtable relation

$$x^n = \sum_{k_1} \sum_{k_2} \cdots \sum_{k_n} C_{k_1} C_{k_2} \cdots C_{k_n} \exp\left[j(\theta_{k_1} + \theta_{k_2} + \cdots + \theta_{k_n})\right] \tag{3.30}$$

Each summation includes all frequencies in both the positive and the negative sense. The resultant set of frequencies contains as a general member

$$\omega_{k_1 k_2 \cdots k_n} = \omega_{k_1} + \omega_{k_2} + \cdots + \omega_{k_n} \tag{3.31}$$

where each term on the right can assume the value of any of the original frequencies in either sense. For the same general frequency the real amplitude and phase angle are given by

$$C_{k_1 k_2 \cdots k_n} = 2C_{k_1} C_{k_2} \cdots C_{k_n}$$

$$= \frac{\text{Im}\,(C_{k_1} C_{k_2} \cdots C_{k_n})}{\text{Re}\,(C_{k_1} C_{k_2} \cdots C_{k_n})} \tag{3.32}$$

Certain features of the output frequencies may be deduced from the above formulas:

1. The initial frequencies are missing in even powers but present in odd powers.

2. A constant is present in even powers but missing in odd powers.

3. For purely harmonic frequencies the harmonic numbers are either all even or all odd in accordance with the power of the nonlinear term.

4. For mixed frequencies the sums of the harmonic numbers are even or odd in accordance with the power of the nonlinear term.

The above analysis and results are restricted to nonlinearities expressed in polynomial form. A more general analysis is available wherein some non-linear characteristics are defined by complex integrals relating the input and output waves. The methods are quite powerful but they are also quite complicated and are of greatest value in the statistical treatment of nonlinear devices. These methods are described in the reports by Bennett, Rice, and Feuerstein which are listed as references for this chapter.

HARMONIC ANALYSIS

The preceding portion of this chapter has shown how a nonlinear function can be represented by a polynomial and how the response of a device to a sinusoidal input can be determined from such a representation. The result is a set of harmonically related vibrations. It frequently happens that the reverse type of analysis is required. The distorted wave is the given function and it is desired to learn the amplitudes and phase angles of its sinusoidal components.

The decomposition of a mathematical function into other functions is usually accomplished by the use of orthogonality. The advantage of this property was clearly demonstrated in the discussion of least-squares approximations. An examination of the definite integrals

$$\int_{-\pi}^{\pi} e^{jmt} e^{-jnt}\, dt = 0 \qquad n \neq m$$

$$= 2\pi \qquad n = m \tag{3.33}$$

shows that harmonically related sinusoidal vibrations form a suitable set of orthogonal functions. If these vibrations are considered to be the components of the given function then the representation must assume the general form

$$f(t) = \sum_{-\infty}^{\infty} c_n e^{jnt} \tag{3.34}$$

A method for evaluating the components is quite apparent. Both sides of the equation are multiplied by one of the orthogonal functions and the result is integrated over the proper interval. Due to orthogonality all terms but one in the infinite series become zero and the result is

$$c_n = \frac{1}{2\pi} \int_{-\pi}^{\pi} f(t) e^{-jnt} \, dt \tag{3.35}$$

The exponential functions can be separated into their sinusoidal components to provide an alternate form

$$f(t) = \frac{a_0}{2} + \sum_{n=1}^{\infty} [a_n \cos nt + b_n \sin nt] \tag{3.36}$$

Due to the even and odd nature of the cosine and sine functions this summation need be performed over only positive integers. Again the orthogonal property of the sinusoids gives the simple formulas for the coefficients

$$a_n = \frac{1}{\pi} \int_{-\pi}^{\pi} f(t) \cos nt \, dt$$

$$b_n = \frac{1}{\pi} \int_{-\pi}^{\pi} f(t) \sin nt \, dt \tag{3.37}$$

Any of the above representations of a given function is known as a *Fourier series*. The amplitudes of the components will be called *Fourier coefficients*. It can immediately be observed that their existence requires only that $f(t)$ is integrable. Since the method for determining these coefficients is so direct it is easy to conclude that the resultant series automatically converges to the proper value. This assumption is quite dangerous and the question of convergence must be thoroughly explored. Unfortunately at this time mathematicians have been unable to produce complete necessary and sufficient conditions on $f(t)$ to assure that its Fourier series is uniformly convergent over the entire interval. This property may in fact be a unique characteristic for classifying functions.

Continuous functions have been found whose Fourier series representations fail at certain points yet other functions which are unbounded and which contain an infinite number of abrupt discontinuities have Fourier series representations that converge everywhere. Acknowledging this lack of clarification and recognizing the wide use of Fourier series it is considered profitable to summarize some of the analytical results which have been achieved at this time. The first part of the discussion presents a few properties of the Fourier coefficients and the second part is concerned with the convergence of the series.

Theorem 3.2   Partial sums using the sum of the squares of corresponding Fourier coefficients form a bounded sequence.

This is proved by considering an integral whose integrand is everywhere positive. It must be true that

$$\int_{-\pi}^{\pi} \left\{ f(t) - \sum_{k=1}^{n} [a_k \cos kt + b_k \sin kt] \right\}^2 dt \geqslant 0 \qquad (3.38)$$

Expanding and utilizing the definitions of a_k and b_k this reduces to

$$\int_{-\pi}^{\pi} [f(t)]^2 \, dt - \pi \sum_{k=1}^{n} (a_k^2 + b_k^2) \geqslant 0 \qquad (3.39)$$

Thus if the function is integrable the above partial sums are bounded for any n. This theorem contains the following corollary as a sufficient condition.

Corollary 3.2.1   The Fourier coefficients form a null sequence.

This means that the harmonic amplitudes approach zero as the harmonic number increases. This corollary forms the basis for the following important result which is due to Riemann and Lebesgue.

Theorem 3.3   Coefficients of the Fourier type which are derived from an integrable function over an interval from a to b which is inside the minimum interval of orthogonality form a null sequence.

This is proved by defining $f(t)$ in terms of the integrable function $\psi(t)$ in the following manner where

$$\begin{aligned} f(t) &= \psi(t) \qquad \text{for} \qquad a \leq t \leq b \\ f(t) &= 0 \qquad \text{elsewhere.} \end{aligned} \qquad (3.40)$$

Since this $f(t)$ is also integrable it is true that

$$\begin{aligned} \alpha_n &= \frac{1}{\pi} \int_a^b \psi(t) \cos nt \, dt \\ &= \frac{1}{\pi} \int_a^b f(t) \cos nt \, dt = a_n \end{aligned} \qquad (3.41)$$

and similarly for $\sin nt$ it follows that $\beta_n = b_n$. Therefore both α_n and β_n approach zero as n becomes sufficiently large. This theorem places no restriction on the locations of a and b within the interval and no limit on how closely they may be spaced.

The above interesting results apply only to the behavior of the Fourier coefficients. They give no assurance that the series converges at any point to the value of $f(t)$. This problem is attacked by expressing a partial sum in closed form. Using several trigonometric identities the sum of terms can be written as

$$
\begin{aligned}
s_n &= \frac{1}{2\pi} \int_{-\pi}^{\pi} f(t)\, dt \\
&\quad + \sum_{k=1}^{n} \left[\frac{1}{\pi} \int_{-\pi}^{\pi} f(u) \cos ku\, du \right] \cos kt \\
&\quad + \sum_{k=1}^{n} \left[\frac{1}{\pi} \int_{-\pi}^{\pi} f(u) \sin ku\, du \right] \sin kt \\
&= \frac{1}{\pi} \int_{-\pi}^{\pi} f(u)[\tfrac{1}{2} + \cos(u-t) + \cdots + \cos n(u-t)]\, du \\
&= \frac{1}{2\pi} \int_{-\pi}^{\pi} f(u)\, \frac{\sin(2n+1)\left(\dfrac{u-t}{2}\right)}{\sin\left(\dfrac{u-t}{2}\right)}\, du \\
&= \frac{1}{2\pi} \int_{-\pi}^{\pi} f(t+u)\, \frac{\sin(2n+1)\dfrac{u}{2}}{\sin\dfrac{u}{2}}\, du \\
&= \frac{1}{\pi} \int_{0}^{\pi} \frac{f(t+u)+f(t-u)}{2}\, \frac{\sin(2n+1)\dfrac{u}{2}}{\sin\dfrac{u}{2}}\, du
\end{aligned}
\tag{3.42}
$$

Changing the variable of integration by a factor of two and making the substitution

$$
\bar{f}(t, u) = \frac{f(t+2u)+f(t-2u)}{2}
\tag{3.43}
$$

permit the partial sum to be expressed in the compact form

$$
s_n(t) = \frac{2}{\pi} \int_{0}^{\pi/2} \bar{f}(t, u)\, \frac{\sin(2n+1)u}{\sin u}\, du
\tag{3.44}
$$

An integral of the above form is called a *Dirichlet integral*. Since it very concisely states the value of n terms of a Fourier series a necessary and sufficient condition for convergence of the series is that the above integral approaches a limit as n increases without bound. Unfortunately this integral is not in a form convenient for such a determination. Since $f(t)$ is assumed to be integrable then the function

$$
\psi(t) = \frac{\bar{f}(t, u)}{\sin u}
\tag{3.45}
$$

is also integrable over regions where the denominator is not zero. The direct application of Theorem 3.3 shows that for any δ inside the interval of integration the integral

$$\frac{2}{\pi} \int_{\delta}^{\pi/2} \psi(t) \sin (2n + 1)u \, du$$

must approach zero as n increases. Since δ can be arbitrarily small all the information regarding the limit of a Fourier series is contained in the behavior of the Dirichlet integral at the origin. Since u is now restricted to small values the approximation of $\sin u$ by u is made. Substituting $m = 2n + 1$ permits Equation 3.44 to be written in the alternate form

$$s_m(t) = \frac{2}{\pi} \int_0^{\delta} \bar{f}(t, u) \frac{\sin mu}{u} \, du \qquad (3.46)$$

This is the basis for Riemann's localization theorem. This puzzling result states that the convergence of a Fourier series depends only on the behavior of $f(t)$ in a small neighborhood around t although the determination of the Fourier coefficients depends on values of $f(t)$ over the entire interval. In Equation 3.46 another requirement on $f(t)$ becomes more apparent. For $s_m(t)$ to be calculable the limit

$$\bar{f}(t, 0) = \lim_{u \to 0} \bar{f}(t, u) \qquad (3.47)$$

must exist for every value of t from $-\pi$ to π. Therefore, necessary and sufficient conditions on a function such that its Fourier series converge are the following:

1. The function is integrable.
2. The function possesses limits on each side of every discontinuity.

The first condition is sometimes stated as the sufficient requirement that the function be bounded and have a finite number of ordinary discontinuities. The second condition is frequently given as a requirement that the function has a finite number of extrema in the interval.

   The above conditions assure convergence of the series but do not establish the value to which the series converges. Numerous rules have been developed for determining this limit. Perhaps the most succinct and practical is contained in a theorem given by Dirichlet in 1829 and illustrated in Figure 3.8.

Theorem 3.4   *If $f(t)$ is monotonic in sufficiently small regions on both sides of every discontinuity then its Fourier series converges everywhere to $f(t)$ except at points of discontinuity where it converges to the average value.*

This is proved by assuming $\bar{f}(t, u)$ to be monotonic in a small region and applying a mean value theorem from integral calculus to give the identity

$$\int_0^\delta \bar{f}(t, u) \frac{\sin mu}{u} \, du = \bar{f}(t, 0) \int_0^{\delta'} \frac{\sin mu}{u} \, du + \bar{f}(t, \delta) \int_{\delta'}^\delta \frac{\sin mu}{u} \, du \quad (3.48)$$

where $0 < \delta' < \delta$. As m increases, the last integral vanishes in accordance with Theorem 3.3. The first integral on the right approaches the definite integral

$$\int_0^\infty \frac{\sin x}{x} \, dx = \frac{\pi}{2} \quad (3.49)$$

Therefore as m increases without bound $s_m(t)$ approaches $\bar{f}(t, 0)$ and the theorem is established.

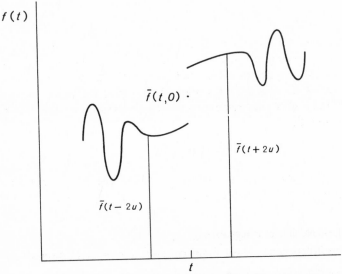

Figure 3.8   A function with a convergent Fourier series.

The integral of an odd function is zero between symmetrical limits. The product of two even or two odd functions is an even function and a cross product of such functions yields an odd function. Reference to Equations 3.37 shows the following:

1. The Fourier series for an even function contains only a constant and cosine terms.

2. The Fourier series for an odd function contains only sine terms.

The expansion for a general function contains both types of terms and is consistent with the interesting fact that any function can be expressed as the sum of an even and an odd function. It is also interesting and sometimes useful to observe that if a wave is symmetrical about a point it is also symmetrical in the same sense about a point separated by one half wavelength.

Regardless of phase angles all harmonic vibrations with even harmonic numbers repeat during the second half of the interval. Vibrations with odd harmonic numbers repeat their wave shapes but have a negative sense. This indicates the following:

1. The Fourier series for a function which is repetitive in a negative sense over both halves of the interval contains only vibrations with odd harmonic numbers.

2. The Fourier series for a function which is repetitive in a positive sense over both halves of the interval contains only vibrations with even harmonic numbers.

These rules may be generalized to include fractions of the interval other than halves. In waves with few harmonic components the graphical addition and subtraction of these portions of the total wave can sometimes be arranged to yield the harmonic amplitudes.

Regardless of the manner in which the original $f(t)$ might be defined outside of the interval from $-\pi$ to π the Fourier series expansion is repetitive with period 2π. If $f(t)$ is defined to be equally repetitive then the evaluation of the Fourier coefficients may be performed over any desired interval with a width of 2π.

It is often desired to expand a function in an interval with another width. This is done by substituting

$$\frac{\tau}{l} = \frac{t}{\pi} \tag{3.50}$$

and defining the expansion in the new form

$$f(\tau) = \frac{a_0}{2} + \sum_n \left[a_n \cos \frac{n\pi\tau}{l} + b_n \sin \frac{n\pi\tau}{l} \right] \tag{3.51}$$

These new functions are orthogonal in the interval from $-l$ to l. Application of the original principle shows that the new coefficients are given by

$$a_n = \frac{1}{l} \int_{-l}^{l} f(\tau) \cos \frac{n\pi\tau}{l} \, d\tau$$

$$b_n = \frac{1}{l} \int_{-l}^{l} f(\tau) \sin \frac{n\pi\tau}{l} \, d\tau \tag{3.52}$$

The formulas given in Equations 3.35, 3.37, and 3.52 require an integration to find the Fourier coefficients. This is usually performed by analytical or numerical methods depending on the manner in which $f(t)$ is specified. The

Fourier coefficients may also be found by graphical methods such as the one shown in Figure 3.9. In this construction $f(t)$ is always plotted along the vertical axes. Either $\sin nt$ or $-\cos nt$ is plotted horizontally. For example, if

$$x = \sin nt$$
$$dx = n \cos nt \, dt$$

(3.53)

then the area in this closed curve is given by

$$\oint f(t) \, dx = n \int_{-\pi}^{\pi} f(t) \cos nt \, dt = n\pi a_n$$

(3.54)

Similarly, if $x = -\cos nt$ then

$$\oint f(t) \, dx = n\pi b_n$$

(3.55)

so the enclosed areas equal the Fourier coefficients weighted by the factor $n\pi$. The symbol on the integral sign means in this case that x passes through enough cycles to close the curve. The construction in Figure 3.9 applies to the second harmonic components and x traverses two complete cycles for each cycle of $f(t)$.

The necessity for calculating or graphically estimating the integrals in Equations 3.37 can be avoided by the use of numerical methods. In one method the interval from $-\pi$ to π is divided into r equal parts and then the integrals which give the Fourier coefficients can be approximated by the

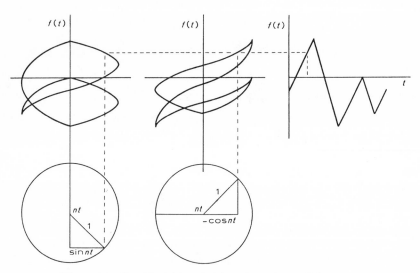

Figure 3.9   Graphical determination of Fourier coefficients.

sums

$$a_n \approx \frac{1}{\pi} \sum_{k=1}^{r} f(t_k) \cos nt_k \, \Delta t$$

$$= \frac{2}{r} \sum_{k=1}^{r} f(t_k) \cos nt_k \tag{3.56}$$

$$b_n \approx \frac{2}{r} \sum_{k=1}^{r} f(t_k) \sin nt_k$$

The samples $f(t_k)$ of the original wave are tabulated against appropriate values of $\cos nt_k$ and $\sin nt_k$. Corresponding entries are multiplied and the results are added to compute a_n and b_n.

A problem with this method is the proper selection of r. This number must remain fixed during the calculation. If r is greater than twice the highest harmonic number which is of interest then enough information is generated to give a unique value for each Fourier coefficient. There is no assurance that these values are correct, however. To give confidence that the results are significant the number r should be greater than twice the highest harmonic number which is present in the Fourier expansion whether or not these higher order harmonic amplitudes are calculated. The method for finding the answer is therefore unfortunately dependent on the nature of the answer.

An alternate method is frequently presented in engineering literature. The interval is again divided into r equal parts and the samples of $f(t)$ are used to establish r simultaneous equations of the form

$$f(t_k) = \frac{a_0}{2} + \sum_{n=1}^{s} \left(a_n \cos nt_k + b_n \sin nt_k \right) \tag{3.57}$$

If $r = 2s + 1$ then there are exactly r undetermined Fourier coefficients and the same number of independent equations. Although this method is frequently used, it contains the same intrinsic difficulty in properly selecting r. Furthermore, if it is desired to determine harmonic amplitudes of low order with a high degree of accuracy it is still necessary to solve simultaneously a large number of equations.

This discussion of Fourier series will be illustrated with a single example which demonstrates rather unexpected behavior. The function to be expanded is shown in Figure 3.10. This is an odd function so it contains only sine terms. It is repetitive in a negative sense over the second half of the interval so it has components with only odd harmonic numbers. The function is described by the relations

$$f(t) = -c \qquad -\pi < t < 0$$
$$f(t) = c \qquad \quad 0 < t < \pi \tag{3.58}$$

and the use of Equations 3.37 shows that its Fourier series is

$$f(t) = \frac{4c}{\pi}\left[\frac{\sin t}{1} + \frac{\sin 3t}{3} + \frac{\sin 5t}{5} + \cdots\right] \qquad (3.59)$$

When five terms of this series are used then the approximation appears as shown in Figure 3.10. The oscillations occur at each corner. If the number of terms is increased the oscillations move toward the points of discontinuity but maintain constant amplitudes. This peculiar phenomenon which was first discussed by Gibbs (1898) will be explained briefly.

Using the notation defined in Equation 3.43 and considering a discontinuity at the origin it is observed that for $t > 0$

$$\bar{f}(t, u) = 0 \qquad u > \frac{t}{2}$$

$$\bar{f}(t, u) = c \qquad u < \frac{t}{2} \qquad (3.60)$$

so the general Dirichlet integral given in Equation 3.46 can be written in the form

$$s_n(t) = \frac{2c}{\pi}\int_0^{t/2} \frac{\sin mu}{u}\, du$$

$$= \frac{2c}{\pi}\int_0^{mt/2} \frac{\sin x}{x}\, dx \qquad (3.61)$$

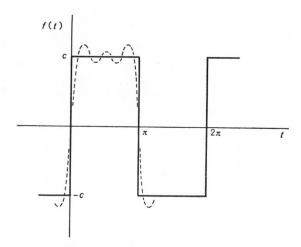

Figure 3.10 The approximation of a square wave by a Fourier series with five terms.

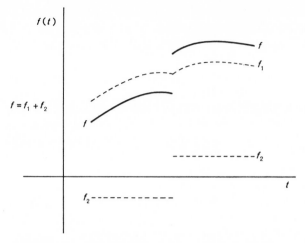

Figure 3.11   Gibbs phenomenon in an arbitrary discontinuous function.

This definite integral appears frequently in analytical work and tabulated values are available. Its evaluation as a function of the upper limit produces the oscillations in Figure 3.10. The nature of the integrand causes maxima at

$$\frac{mt}{2} = \pi, 3\pi, 5\pi, \dots \tag{3.62}$$

and minima at

$$\frac{mt}{2} = 2\pi, 4\pi, 6\pi, \dots \tag{3.63}$$

The first pair are extreme values and introduce the factors $1.1790\cdots$ and $0.9022\cdots$ relative to half the amplitude of the discontinuity. These factors are obviously independent of m. It is apparent from Equations 3.62 and 3.63 that an increase in m only decreases the values of t at which extrema occur.

This phenomenon can be regarded as a defect in the Fourier series representation and it is appropriate to ask when it will occur, to what extent, and how it can be eliminated. The first two questions are answered easily. Figure 3.11 shows how a discontinuous function f can be separated into two functions which have special properties. The function f_1 is continuous and its Fourier series is convergent if it meets the requirements of Theorem 3.4. The function f_2 has the form of the function shown in Figure 3.10 so it displays the Gibbs phenomenon. The exact nature of the effect can be calculated using Equation 3.61 with an appropriate value for c. The separation may be performed a finite number of times in the interval and the Gibbs phenomenon will occur each time.

The Gibbs phenomenon at first appears to contradict Theorem 3.4. However, for any t it is always possible to use enough terms in the series to place the oscillations so close to the discontinuity that the series converges at the selected point.

The elimination of the Gibbs phenomenon requires a rather elegant technique. It is somewhat astonishing to learn that the method is based on a procedure for achieving convergence of a normally divergent series. Starting at least as early as the work of Euler in 1755 various mathematicians have proposed such methods. It is a requirement of these techniques that the new series converges to the same value as the old series when the original series converges. These methods either operate in some manner on each term of the original series or else form partial sums in different fashions. Some of the methods bear the names of Abel, Borel, Cesàro, Hölder, Le Roy, and Riesz. These are described in detail in mathematics literature such as the book by Knopp which is listed as a reference for this chapter.

In 1904 Fejér showed that the application of one of these techniques to a Fourier series produces a new series which is more strongly convergent. It also lacks the Gibbs phenomenon. Fejér used a procedure usually attributed to Cesàro (1890) although it is identical in the first order to a method proposed in 1882 by Hölder.

In the first order Cesàro procedure each partial sum is defined as the arithmetic mean of all prior partial sums in the original series. In forming a Fejér series from a Fourier series

$$\sigma_n(t) = \frac{s_0(t) + s_1(t) + \cdots + s_{n-1}(t)}{n} \tag{3.64}$$

Since each Fourier partial sum is given by

$$s_n(t) = \frac{2}{\pi} \int_0^{\pi/2} \bar{f}(t, u) \frac{\sin (2n + 1)u}{\sin u} \, du \tag{3.65}$$

the substitution of the trigonometric identity

$$\sum_{k=0}^{n-1} \sin (2k + 1) = \frac{\sin^2 nu}{\sin u} \tag{3.66}$$

permits a Fejér partial sum to be expressed as

$$\sigma_n(t) = \frac{2}{n\pi} \int_0^{\pi/2} \bar{f}(t, u) \left(\frac{\sin nu}{\sin u} \right)^2 du \tag{3.67}$$

Comparing this expression with Equation 3.44 which gives the Fourier partial sums it is observed that the Fejér integral is divided by n and it contains a weighting factor which is always positive. These two results produce the improved convergence properties of the Fejér series.

Substitution of the trigonometric identity

$$\frac{\sin (2n + 1)u}{\sin u} = 1 + \sum_{k=0}^{n-1} 2 \cos 2(k + 1)u \qquad (3.68)$$

into Equation 3.66 provides a means for evaluating the definite integral

$$\frac{2}{n\pi} \int_0^{\pi/2} \left(\frac{\sin nu}{\sin u}\right)^2 du = 1 \qquad (3.69)$$

This relation is multiplied by $\bar{f}(t, 0)$ and subtracted from the Fejér integral to give the error expression

$$\sigma_n(t) - \bar{f}(t, 0) = \frac{2}{n\pi} \int_0^{\pi/2} \phi(t, u)\left(\frac{\sin nu}{\sin u}\right)^2 du \qquad (3.70)$$

where the substitution

$$\phi(t, u) = \bar{f}(t, u) - \bar{f}(t, 0) \qquad (3.71)$$

is made to simplify notation in further equations. For the Fejér series to converge properly it is necessary and sufficient that the above integral approaches zero as n increases. The expression can fortunately be replaced with several equivalent forms which are more convenient. Within the limits of integration

$$\left(\frac{\sin nu}{\sin u}\right)^2 - \left(\frac{\sin nu}{u}\right)^2 < K \qquad (3.72)$$

where K is constant. The difference between the integrals

$$\left|\frac{2}{n\pi} \int_0^{\pi/2} \phi(t, u)\left(\frac{\sin nu}{\sin u}\right)^2 du\right| - \left|\frac{2}{n\pi} \int_0^{\pi/2} \phi(t, u)\left(\frac{\sin nu}{u}\right)^2 du\right|$$

$$\leqslant \frac{2K}{n\pi} \int_0^{\pi/2} |\phi(t, u)| \, du \qquad (3.73)$$

vanishes as n increases since $\phi(t, u)$ must also be integrable. Therefore an equivalent necessary and sufficient condition for proper convergence of a Fejér series is

$$\lim_{n \to \infty} \frac{2}{n\pi} \int_0^{\pi/2} \phi(t, u)\left(\frac{\sin nu}{u}\right)^2 du = 0 \qquad (3.74)$$

For any u and δ within the limits of integration which satisfy $u > \delta > 0$ it is true that

$$\left(\frac{\sin nu}{u}\right)^2 \leqslant \frac{1}{u^2} \leqslant \frac{1}{\delta^2} \qquad (3.75)$$

This establishes the inequality

$$\left| \frac{2}{n\pi} \int_{\delta}^{\pi/2} \phi(t, u) \left(\frac{\sin nu}{u} \right)^2 du \right| \leqslant \frac{2}{n\pi\delta^2} \int_{\delta}^{\pi/2} |\phi(t, u)| \, du \qquad (3.76)$$

where the expression on the right vanishes as n increases. Therefore the essential information about Fejér series convergence is contained in the interval from 0 to δ. Since δ can be arbitrarily small these convergence properties again depend on the local behavior of $f(t)$. An alternate requirement for convergence is

$$\frac{2}{n\pi} \int_0^{\delta} \phi(t, u) \left(\frac{\sin nu}{u} \right)^2 du \to 0 \qquad (3.77)$$

The series converges to $\bar{f}(t, 0)$ if the expression approaches zero either as n increases or as δ decreases.

Theorem 3.5 *The Fejér series for $f(t)$ converges to $\bar{f}(t, 0)$ wherever this limit exists.*

This theorem is established by observing that wherever $\bar{f}(t, 0)$ exists it is possible to select δ such that for the given t and for all u satisfying $0 \leq u \leq \delta$

$$|\phi(t, u)| = |\bar{f}(t, u) - \bar{f}(t, 0)| < \epsilon \qquad (3.78)$$

so it is true that

$$\left| \frac{2}{n\pi} \int_0^{\delta} \phi(t, u) \left(\frac{\sin nu}{u} \right)^2 du \right| \leqslant \frac{2\epsilon}{n\pi} \int_0^{\pi/2} \left(\frac{\sin nu}{u} \right)^2 du = \epsilon \qquad (3.79)$$

and the series must converge. It should be observed that the above extension of the limit of integration requires a positive integrand to maintain the inequality. This procedure is not possible in the corresponding expression for a Fourier series since the weighting factor is oscillatory. Therefore there is no counterpart to Theorem 3.5 in the theory of Fourier series.

Corollary 3.5.1 *At a point of ordinary discontinuity the Fejér series converges to the average value of the function.*

This follows immediately from the theorem.

Corollary 3.5.2 *Wherever $f(t)$ is continuous its Fejér series converges uniformly.*

This is also a consequence of Theorem 3.5 since the limit $\bar{f}(t, 0)$ exists for all values of t where $f(t)$ is continuous. Again the corresponding statement is untrue for Fourier series.

To complete the comparison between Fejér and Fourier series the manner in which each represents a square wave is shown in Figure 3.12. The Fourier

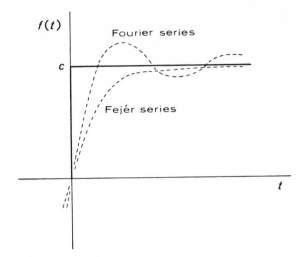

Figure 3.12   Comparison of Fourier and Fejér series.

series curve is a graph of

$$\frac{2c}{\pi} \int_0^{mt/2} \frac{\sin x}{x}\, dx$$

as shown previously. The Fejér series representation is a plot of

$$\frac{2c}{\pi} \int_0^{nt/2} \left(\frac{\sin x}{x}\right)^2 dx$$

which is derived from Equation 3.70 by letting $\bar{f}(t, 0) = 0$ and observing that $\bar{f}(t, u) = 0$ when $2u > t$. The presence of the positive integrand in the Fejér expression prohibits the oscillatory behavior known as the Gibbs phenomenon.

When the Fejér coefficients are calculated from Equation 3.64 it is observed that they are smaller than the original Fourier coefficients in every case except for the constant term and those coefficients which are zero. The constant term in the Fejér expansion again indicates the average value of the function. The fact that the Fejér and Fourier expansions differ in every oscillatory term is of considerable importance because it indicates that trigonometric interpolation does not yield a unique result. It may be some-what astonishing to realize that it is incorrect to consider that a given wave possesses a harmonic component of a certain amplitude unless the approx-imation process is also stipulated. The possession of a certain harmonic component is not an intrinsic property of a given wave as is commonly believed. It is also observed from Equation 3.64 that the lower order Fejér coefficients approach their Fourier counterparts as the number of terms in the series increases. This creates the suspicion that the difference in behavior between Fejér and Fourier series can be attributed primarily to the higher order terms.

Theorem 3.1 due to Weierstrass was stated early in this chapter without proof. Corollary 3.5.2 provides a means for establishing this theorem. A change of variable will convert $F(x)$ to an $f(t)$ having an interval of continuity no greater than 2π. For this function there exists an m such that

$$|f(t) - \sigma_m(t)| < \frac{\epsilon}{2} \tag{3.80}$$

for all values of t in the interval. The finite number of trigonometric functions contained in $\sigma_m(t)$ can be expressed to any degree of accuracy by a uniformly convergent power series. Therefore there exists an n such that

$$|p_n(t) - \sigma_m(t)| < \frac{\epsilon}{2} \tag{3.81}$$

These can be combined to give

$$|f(t) - p_n(t)| < \epsilon \tag{3.82}$$

and the theorem is proved. As a matter of logic it should be observed that the Weierstrass theorem was not used to develop the convergence properties of Fejér series.

DIRECT HARMONIC PREDICTION

Sometimes it is desired to estimate harmonic generation directly from the nonlinear characteristic. Figure 3.13 shows how this may be accomplished graphically. It is an extension of the method shown in Figure 3.9. The circle on the right is used to generate an input $x = A \cos \theta$. The remainder of the procedure is the same as the previous method. The technique is completely general and gives proper values when the nonlinearity contains hysteresis as shown.

The harmonic amplitudes can also be found by a method presented by Lewis in 1955. In this method it is assumed that a signal $x = \cos \theta$ is introduced to the nonlinear device. The output is considered to be a direct function of θ and is therefore represented as a Fourier series of the form

$$y(x) = y(\cos \theta) = \frac{a_0}{2} + \sum_n (a_n \cos n\theta + b_n \sin n\theta) \tag{3.83}$$

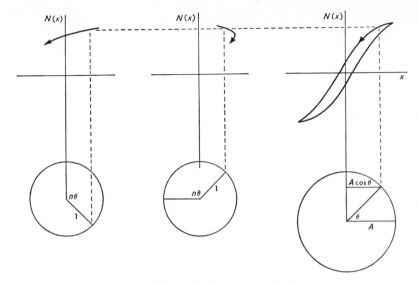

Figure 3.13 Direct harmonic prediction.

where the coefficients are given by

$$a_n = \frac{1}{\pi} \int_{-\pi}^{\pi} y(\cos \theta) \cos n\theta \, d\theta$$

$$b_n = \frac{1}{\pi} \int_{-\pi}^{\pi} y(\cos \theta) \sin n\theta \, d\theta$$

(3.84)

Using the relation

$$dx = -\sin \theta \, d\theta = -\sqrt{1 - x^2} \, d\theta$$

(3.85)

the variable of integration may be changed and the integrals assume the form

$$a_n = -\frac{1}{\pi} \oint y(x) \cos (n \cos^{-1} x) \frac{dx}{\sqrt{1 - x^2}}$$

$$b_n = -\frac{1}{\pi} \oint y(x) \sin (n \cos^{-1} x) \frac{dx}{\sqrt{1 - x^2}}$$

(3.86)

The integral symbol again signifies the fact that the integrations are performed in the x domain so as to correspond to an entire cycle for θ. This generality is necessary if $y(x)$ contains an unsymmetric hysteresis loop.

A portion of the integrands may be recognized as the definitions frequently used for a remarkable set of orthogonal algebraic expressions known as

Chebyshev polynomials. These definitions are

$$T_n(x) = \cos n\theta = \cos (n \cos^{-1} x)$$
$$U_n(x) = \sin n\theta = \sin (n \cos^{-1} x)$$

(3.87)

From the trigonometric identities

$$\sin^2 n\theta + \cos^2 n\theta = 1$$
$$2 \cos \theta \cos n\theta = \cos (n + 1)\theta + \cos (n - 1)\theta$$
$$2 \cos \theta \sin n\theta = \sin (n + 1)\theta + \sin (n - 1)\theta$$

(3.88)

there are derived immediately several useful relations between these functions

$$T_n^2 + U_n^2 = 1$$
$$2xT_n = T_{n+1} + T_{n-1}$$
$$2xU_n = U_{n+1} + U_{n-1}$$

(3.89)

Using these relations in addition to the original definitions for T_n and U_n it is possible to generate both sets of algebraic expressions. The first five members of each of these sets are given in the table below.

n	T_n	U_n
0	1	0
1	x	$\sqrt{1 - x^2}$
2	$2x^2 - 1$	$2x\sqrt{1 - x^2}$
3	$4x^3 - 3x$	$(4x^2 - 1)\sqrt{1 - x^2}$
4	$8x^4 - 8x^2 + 1$	$(8x^3 - 4x)\sqrt{1 - x^2}$

It is observed that $U_1(x)$ is identical to the function appearing in the denominator of the integral expressions for a_n and b_n. This permits the harmonic amplitudes to be expressed in the very interesting and compact forms

$$a_n = -\frac{1}{\pi} \oint y(x) \frac{T_n(x)}{U_1(x)} dx$$
$$b_n = -\frac{1}{\pi} \oint y(x) \frac{U_n(x)}{U_1(x)} dx$$

(3.90)

These integrals yield the harmonic amplitudes directly using only the non-linear characteristic $y(x)$ weighted with Chebyshev polynomials. The integration lies between the extremes of $+1$ and -1 in the x domain. If the amplitude of the input signal is different or if there is a bias to this signal then a new $y(x)$ must be constructed which meets these requirements. This is not a difficult procedure, however, since it involves only an appropriate change of variable.

The function $U_1(x)$ becomes zero at the extreme values for x. This introduces the most serious problem in the application of this method. Satisfactory results in numerical integrations may be achieved by assuming that the integrand has constant slope in a region sufficiently close to the singular points.

If the nonlinear characteristic is single valued then all b_n are zero. If the characteristic is double valued but symmetrical about the origin then some b_n are not zero. In symmetrical cases the integration may be performed for a half cycle in the θ domain. If the nonlinear characteristic is defined in graphical or tabular form then the a_n and b_n must be determined by numerical integration using tabulated values for the Chebyshev polynomials.

After the coefficients have been found it is possible to combine Equations 3.83 and 3.87 to express $y(x)$ in the algebraic form

$$y(x) = \frac{a_0}{2} + \sum_n [a_n T_n(x) + b_n U_n(x)] \qquad (3.91)$$

Although this method is presented as a means of harmonic analysis it also yields an algebraic approximation for the nonlinear characteristic. Due to the relations between the variables x and θ the original expansion of y (cos θ) as a Fourier series is equivalent to an expansion of $y(x)$ using the orthogonal Chebyshev polynomials. It is quite interesting to observe that the constants a_n and b_n are the coefficients of cos $n\theta$ and sin $n\theta$ in the frequency domain and also of $T_n(x)$ and $U_n(x)$ in the displacement domain.

REFERENCES

1. F. Willers, *Practical Analysis*, Dover, New York, 1948.
2. F. Hildebrand, *Introduction to Numerical Analysis*, McGraw-Hill, New York, 1956.
3. G. Sansone, *Orthogonal Functions*, Interscience Publishers, New York, 1959.
4. K. Knopp, *Theory and Application of Infinite Series*, Hafner, New York, 1947.
5. R. Manley, *Waveform Analysis*, Chapman and Hall, London, 1950.
6. W. R. Bennett, New Results in the Calculation of Modulation Products, *Bell System Technical Journal*, Volume 12, 1933, pp. 228–243.
7. E. Feuerstein, Intermodulation Products for Nu-Law Biased Wave Rectifier for Multiple Frequency Inputs, *Quarterly Journal of Applied Mathematics*, Volume 15, 1957, pp. 183–192.
8. S. Rice, Mathematical Analysis of Random Noise, *Bell System Technical Journal*, Volume 24, 1945, pp. 52–162.
9. L. J. Lewis, Harmonic Analysis for Nonlinear Characteristics, *Transactions of the AIEE*, Volume 74, 1955, pp. 693–700.

4

THE POWER FORMULAS

Nonlinear processes generate new frequencies and suppress existing ones. It might appear that the infinity of variations in nonlinearities would permit the output frequencies to be generated in any desired proportions. This is not true. For each basic type of device there are relations which must be satisfied by the power flowing into and out of the device at all of the different frequencies which are present. These formulas resemble the expression for conservation of energy but the power present at each frequency is weighted by an added factor. The relations appear as both equalities and inequalities. To illustrate their form one of the more important formulas will be derived from very basic considerations.

The frequencies of the primary sources or generators are called *base frequencies* and each corresponding angular velocity is designated ω_b. Given a set of incommensurable base frequencies the angular velocity of any member of the total ensemble of frequencies which are present in the device may be expressed uniquely by the relation

$$\omega_k = \sum_b M_{kb}\omega_b \qquad (4.1)$$

where M_{kb} is the harmonic number for that portion of ω_b appearing in ω_k. In the mixing process M_{kb} is always rational and is usually a positive or negative integer or zero. Different sets of frequencies are usually present in the input and output of the device so to assure completeness a summation over n will include *all* frequencies which are present. Assuming only that energy is conserved in the device and then interchanging the order of summations provide the relationships

$$0 = \sum_n P_n$$
$$= \sum_n \frac{P_n}{\omega_n}\omega_n$$

$$= \sum_n \frac{P_n}{\omega_n} \sum_b M_{nb}\omega_b$$

$$= \sum_b \omega_b \sum_n \frac{P_n M_{nb}}{\omega_n} \qquad (4.2)$$

Since all of the ω_b are independent this expression can be identically zero only if each of the coefficients is zero. Therefore for *each* base frequency there is a unique relation of the form

$$\sum_n \frac{M_{nb}}{\omega_n} P_n = 0 \qquad (4.3)$$

This is one set of the power formulas. The derivation shows only that the principle of conservation of energy yields a set of simultaneous linear relations which must be satisfied by the power flow at all of the frequencies present. While conservation of energy is necessary for the above result it is not sufficient since other mathematical procedures give different formulas. Also, the method shown above does not indicate the type of dissipationless device for which the resultant formula is applicable. Further analysis will show that it applies only to devices wherein power is proportional to frequency. Such primitive derivations are obviously futile so a more precise development is given in the following pages. Dissipative devices are also included. Alternate proofs can be found in the reports of the original investigators and in the book by Penfield. These are listed as references for this chapter.

TIME AVERAGES

Since a nonlinear device sometimes suppresses input frequencies, sometimes generates new frequencies, and always combines input frequencies it is apparent that the expression of its performance as time dependent functions may lead to extremely complicated results. Analysis is considerably simplified by the use of time averages. This method yields expressions involving

$$\langle f(t) \rangle = \lim_{T \to \infty} \frac{1}{T} \int_0^T f(t)\, dt \qquad (4.4)$$

If $f(t)$ is oscillatory about a zero mean value then its time average is zero. The product of two sinusoids of different frequencies forms such an oscillatory function. However, if the waves have the same frequency and are not in perfect quadrature then the time average is not zero. Therefore the formation of a time average is somewhat analogous to the application of orthogonality to sinusoids except that the limit of integration is not fixed and the frequencies need not be harmonically related.

Several times in the following discussion it will be necessary to use the basic time averages

$$\langle \cos(\theta + \Phi) \cos(\theta + \Psi) \rangle = \tfrac{1}{2} \cos(\Phi - \Psi)$$
$$\langle \sin(\theta + \Phi) \cos(\theta + \Psi) \rangle = \tfrac{1}{2} \sin(\Phi - \Psi) \qquad (4.5)$$
$$\langle \sin(\theta + \Phi) \sin(\theta + \Psi) \rangle = \tfrac{1}{2} \cos(\Phi - \Psi)$$

where Φ and Ψ are arbitrary phase angles and $\theta = \omega t$. The above relations are immediately established by using trigonometric identities which decompose the products into sums of several terms and then discarding the terms involving only θ since such terms are oscillatory about a zero mean.

Theorem 4.1   *The time average of the product of two composite waves derived from the same base frequencies is stationary against changes in phase or frequency of the generators.*

This theorem will be demonstrated by expressing the two waves in the form

$$x = \sum_r X_r \cos(\theta_r + \Phi_r)$$
$$y = \sum_s Y_s \cos(\theta_s + \Psi_s) \qquad (4.6)$$

where in each case $\theta_n = \omega_n t$ and the coefficients are constant. Applying Equations 4.5 gives the result

$$\langle xy \rangle = \tfrac{1}{2} \sum_n X_n Y_n \cos(\Phi_n - \Psi_n) \qquad (4.7)$$

After a phase shift $\Delta\theta_b$ in one of the generators the two waves can be expressed in the form

$$x + \Delta x = \sum_r X_r \cos(\theta_r + \Delta\theta_r + \Phi_r)$$
$$y + \Delta y = \sum_s Y_s \cos(\theta_s + \Delta\theta_s + \Psi_s) \qquad (4.8)$$

From Equation 4.1 it is apparent that the phase shift at any dependent frequency is given by

$$\Delta\theta_n = M_{nb}\, \Delta\theta_b \qquad (4.9)$$

Therefore for equal frequencies in x and y it is true that $\Delta\theta_r = \Delta\theta_s$. Equations 4.8 then have the same form as Equations 4.6 and the time average of the product again is

$$\langle (x + \Delta x)(y + \Delta y) \rangle = \tfrac{1}{2} \sum_n X_n Y_n \cos(\Phi_n - \Psi_n) \qquad (4.10)$$

This establishes the stationary properties of the time average against a change in phase of a single generator. The theorem also holds when the phases of

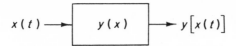

Figure 4.1   The input and output functions.

several generators are shifted simultaneously because $\Delta\theta_n$ again has the same value in the x and y waves.

The stationary property of the time average against a change in frequency is easily demonstrated by repeating the above proof with the substitutions $\theta_n = \omega_n t$ and assuming a frequency shift $\Delta\omega_b$.

> **Theorem 4.2**   *When the input to a nonlinear device is a composite wave derived from several base frequencies the time average of the product of the input and output waves is stationary against changes in phase and frequency of the generators if all other variables remain constant.*

This may appear to be an alternate statement of Theorem 4.1 but there is a significant distinction. The previous theorem assumed that each amplitude remained constant under the change in generator phase or frequency. Such an assumption cannot be made when x and y represent the input and output of a nonlinear device. The characteristics of some devices are dependent on frequency and it is not immediately apparent whether or not such dependency could invalidate the theorem.

To demonstrate this theorem the input wave will be written as

$$x = \sum_r X_r \cos (\theta_r + \Phi_r) \tag{4.11}$$

where all X_r and Φ_r are arbitrary and each θ_r is derived from the base frequencies. The output wave is generated as shown in Figure 4.1 and must be written in a general form such as

$$y[x(t)] = \sum_s [A_s \cos \theta_s + B_s \sin \theta_s] \tag{4.12}$$

since there is no immediate way to calculate either the amplitude or phase angle at each frequency. The form of Equation 4.12 suggests the application of Fourier analysis to determine the coefficients. For this to be possible the output wave must be of such form that its Fourier series converges. It is interesting to observe that since $x(t)$ is well behaved the requirements on the nonlinear characteristic and the output wave are the same. Numerous sufficient conditions for convergence were discussed in the preceding chapter and the applicability of the following proof and perhaps the validity of the theorem itself depend on such conditions being satisfied.

Due to the fact that a nonlinear process always yields frequencies which are integrally related to the input frequencies the entire ensemble of output frequencies has some characteristics of an orthogonal set. An appropriate ensemble of base frequencies must be used simultaneously, however, to provide such orthogonality. This gives a means for determining the Fourier coefficients in the more general case when the function is derived from several incommensurable base frequencies. The calculation of the coefficients in this case may be indicated symbolically by

$$A_s = \frac{2}{(2\pi)^m} \int_{-\pi}^{\pi} \cdots \int_{-\pi}^{\pi} y(x) \cos \theta_s \, d\theta_s$$

$$B_s = \frac{2}{(2\pi)^m} \int_{-\pi}^{\pi} \cdots \int_{-\pi}^{\pi} y(x) \sin \theta_s \, d\theta_s$$

(4.13)

where m is the number of base frequencies present in each θ_s. There are m integrations and the indicated element of integration $d\theta_s$ contains m separate elements, one for each base frequency present.

The output wave is simultaneously a function of the input frequencies ω_r, the input amplitudes X_r, the input phase angles Φ_r, and the nonlinear characteristic $y(x)$. Since the angles associated with the base frequencies are the variables of integration in the above integrals the coefficients A_s and B_s are not functions of these frequencies. They depend on only the input amplitudes, the input phase angles, and the nonlinear characteristic. They are therefore stationary against changes in the base frequencies. Since the other variables are assumed to be constant the time average of the product of the input and output waves must also be stationary against changes in the generator frequencies.

The stationary property against changes in the phase angles of the primary sources follows immediately. A change in any θ_b would have no effect on the above integrals.

It should be observed that the theorem specifically does *not* apply to shifts in the individual phase angles Φ_r of the input wave. For example, if a portion of the input signal is a square wave derived from a particular generator then the time average is independent of the phase angle or frequency of that generator. It is not necessarily stationary against changes in the relative phase angles of the components of the square wave since such changes would immediately alter the shape of the wave.

Corollary 4.2.1 *The amplitude and phase angle of each component of the output wave in a nonlinear device are stationary against changes in phase and frequency of the generators if all other variables remain constant.*

This fact was established during the proof of the theorem.

GENERAL FORMULAS

In a real network the power is a function of the voltage and current in an electrical device or the force and velocity in a mechanical device. To derive formulas for all of the known types of network elements involves highly repetitious mathematical procedures. This will be avoided by developing a set of formulas in canonical form using generalized variables and later applying these results to specific devices.

The formulas are derived from a consideration of various time averages which may be calculated when the phase of one of the generators is arbitrarily shifted. The theoretical justification for this is based partly on Theorem 4.2. Figure 4.2 shows in normalized form that for each frequency present in a composite wave the displacement created by the generator phase shift has components which are both in phase and in quadrature with the original wave. The negative signs appear because the original wave is assumed to be a cosine function with angle $(\theta_r + \Phi_r)$. This same result can be achieved using analytical methods involving trigonometric identities.

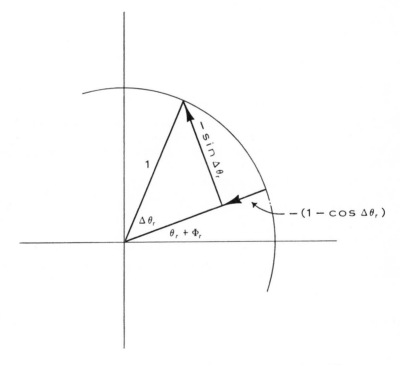

Figure 4.2   Components created by a generator phase shift.

It is possible to separate the displacement in the composite wave into two ensembles, those in phase with their counterparts and those in quadrature with them. For the two composite waves x and y these ensembles can be expressed as

$$\Delta x_p = -\sum_r X_r \cos(\theta_r + \Phi_r)(1 - \cos \Delta\theta_r)$$

$$\Delta x_q = -\sum_r X_r \sin(\theta_r + \Phi_r)(\sin \Delta\theta_r)$$

$$\Delta y_p = -\sum_s Y_s \cos(\theta_s + \Psi'_s)(1 - \cos \Delta\theta_s)$$ (4.14)

$$\Delta y_q = -\sum_s Y_s \sin(\theta_s + \Psi'_s)(\sin \Delta\theta_s)$$

The total displacements are given by

$$\Delta x = \Delta x_p + \Delta x_q$$
$$\Delta y = \Delta y_p + \Delta y_q$$ (4.15)

Using Equations 4.5 to express the time averages of the products of quantities appearing in Equations 4.6 and 4.14 immediately gives

$$\sum_n X_n Y_n \cos(\Phi_n - \Psi'_n)(1 - \cos \Delta\theta_n) = -\langle y \, \Delta x_p \rangle$$

$$\sum_n X_n Y_n \sin(\Phi_n - \Psi'_n)(\sin \Delta\theta_n) = -\langle y \, \Delta x_q \rangle$$ (4.16)

These are the generalized formulas in a canonical form. All of the various power formulas can be derived directly from these equations. The method consists of relating the left side of the equations to any form of power existing in the device and determining those characteristics of x and y which provide an evaluation of the time averages which appear on the right. Unfortunately a complete set of necessary and sufficient conditions for these evaluations has not been discovered. Various sufficient conditions are given as a set of lemmas.

Lemma 4.1 *If x possesses odd symmetry at some point in time and if $y(x)$ is an even function then $\langle y \, \Delta x_p \rangle = 0$.*

Since x has odd symmetry at a known point it is possible to introduce a translation in time such that x has the form

$$x = \sum_r X_r \sin \theta_r$$ (4.17)

The ensemble Δx_p must also have odd symmetry and can be written in the similar form

$$\Delta x_p = \sum_r X'_r \sin \theta_r$$ (4.18)

If $y(x)$ is an even function then the output wave is also even and the time average

$$\lim_{T \to \infty} \frac{1}{T} \int_0^T y \, \Delta x_p \, dt \tag{4.19}$$

is zero. The calculation of the time average about some remote point in time gives valid results since the resultant formula is considered to be applicable only to steady-state conditions. The requirement for odd symmetry in x will be met whenever all of its components are zero simultaneously. The construction of such a situation can sometimes be achieved by arbitrarily shifting the phase angles of the generators. However, the requirement that $y(x)$ have even symmetry is found in few practical devices so the lemma has limited applicability. The following evaluation is much more useful since it places very limited demands on x and the required form for $y(x)$ is found frequently.

Lemma 4.2 *If the slope of a single-valued function $y(x)$ is positive or zero everywhere within the interval and if Δy is not always zero then $\langle y \, \Delta x_p \rangle$ is negative.*

Using the identities of Equation 4.5 to evaluate the time averages of the products of quantities defined in Equation 4.15 gives the unusual result

$$-\langle y \, \Delta x_p \rangle = \langle \Delta y \, \Delta x \rangle \tag{4.20}$$

Since the slope of $y(x)$ is everywhere positive or zero then whenever Δy is not zero it has the same sign as Δx. The time average of the product is necessarily positive and the lemma is established.

Lemma 4.3 *If x possesses odd symmetry at some point in time and if $y(x)$ is an odd function then $\langle y \, \Delta x_q \rangle = 0$.*

This is similar to Lemma 4.1 except that the quadrature component of the displacement is an even function and the output wave is odd. The time average of the product is necessarily zero.

Lemma 4.4 *If x possesses even symmetry at some point in time and if $y(x)$ is single valued then $\langle y \, \Delta x_q \rangle = 0$.*

It is interesting to observe the fact that $y(x)$ need be only single valued for the output wave to be an even function. Since the quadrature component of the displacement is odd the time average of the product is zero.

Lemma 4.5 *If each component of the output wave is in phase with the component of equal frequency in the input wave then $\langle y \, \Delta x_q \rangle = 0$.*

Since each $\Psi_n = \Phi_n$ the quadrature component of the displacement at each frequency is also in quadrature with the component of equal frequency in y.

Since the time average at each frequency is zero the time average of the ensemble is also zero. A sufficient condition for this situation is as follows:

1. The components of the input signal are not shifted in phase by the nonlinear process.

2. Newly generated signals are in phase with similar signals in the input should they be present.

The first part of this condition is met if $y(x)$ is an odd function. This lemma is therefore related to Lemma 4.3. To establish the applicability of the second part of the condition requires a knowledge of the exact nature of $y(x)$ and the components of x.

The formulas stated in Equations 4.16 assume an entirely different form when the arbitrary displacement in x is infinitesimal. Using series expansions for the trigonometric functions and retaining only the first and second order terms give the relations

$$\sum_n X_n Y_n \cos{(\Phi_n - \Psi_n)}(\delta\theta_n)^2 = -\langle y\, \delta x_p\rangle$$

$$\sum_n X_n Y_n \sin{(\Phi_n - \Psi_n)}(\delta\theta_n) = -\langle y\, \delta x_q\rangle$$

$$(4.21)$$

For infinitesimal displacements Equation 4.9 has the form

$$\delta\theta_n = M_{nb}\, \delta\theta_b \qquad (4.22)$$

Incorporating the last relation gives an alternate set of canonical equations

$$\sum_n X_n Y_n \cos{(\Phi_n - \Psi_n)}M_{nb}{}^2 = -\frac{1}{(\delta\theta_b)^2}\langle y\, \delta x_p\rangle$$

$$\sum_n X_n Y_n \sin{(\Phi_n - \Psi_n)}M_{nb} = -\frac{1}{\delta\theta_b}\langle y\, \delta x_q\rangle$$

$$(4.23)$$

The evaluation of the terms on the right which involve infinitesimal displacements is more definitive than the determination with finite displacements. A specific evaluation can be made for each. Using the form of Equation 4.20 which applies to an infinitesimal displacement gives

$$-\frac{1}{(\delta\theta_b)^2}\langle y\, \delta x_p\rangle = \frac{1}{(\delta\theta_b)^2}\langle \delta y\, \delta x\rangle$$

$$= \frac{1}{(\delta\theta_b)^2}\lim_{T\to\infty}\frac{1}{T}\int_0^T \left(\frac{\partial y}{\partial\theta_b}\delta\theta_b\right)\left(\frac{\partial x}{\partial\theta_b}\delta\theta_b\right) dt$$

$$= \lim_{T\to\infty}\frac{1}{T}\int_0^T \frac{\partial y}{\partial\theta_b}\frac{\partial x}{\partial\theta_b} dt$$

$$= \lim_{T\to\infty}\frac{1}{T}\int_0^T \frac{\partial y}{\partial x}\left(\frac{\partial x}{\partial\theta_b}\right)^2 dt \qquad (4.24)$$

This is necessarily positive if the slope of $y(x)$ is everywhere positive or zero and if at some time the input signal passes through a region of positive slope. This result corresponds directly to Lemma 4.2.

It is apparent from Equation 4.14 that δx_p is a second-order term and may be considered to be zero in a first-order analysis. As a consequence

$$
\begin{aligned}
- \frac{1}{\delta\theta_b} \langle y\,\delta x_q \rangle &= - \frac{1}{\delta\theta_b} \langle y\,\delta x \rangle \\
&= - \frac{1}{\delta\theta_b} \lim_{T\to\infty} \frac{1}{T} \int_0^T y\!\left(\frac{\partial x}{\partial\theta_b}\,\delta\theta_b \right) dt \\
&= - \lim_{T\to\infty} \frac{1}{T} \int_0^T y\!\left(\frac{\partial x}{\partial\theta_b}\,dt \right) \\
&= - \lim_{T\to\infty} \frac{1}{\omega_b T} \int_0^{t=T} y\,dx
\end{aligned}
\tag{4.25}
$$

If $y(x)$ is single valued then the last integral is oscillatory about a zero mean value and the time average is zero. This is also true when x has a constant term. This result corresponds somewhat to Lemma 4.4 but there is no longer any requirement on the behavior of x.

Assuming that $y(x)$ is single valued when required, that its slope is positive or zero when required, or that one of the null conditions of Lemmas 4.3, 4.4, or 4.5 is met should the corresponding formula be used, then all of the above results can be summarized as

$$
\begin{aligned}
\sum_n X_n Y_n \cos(\Phi_n - \Psi_n)(1 - \cos M_{nb}\,\Delta\theta_b) &> 0 \\
\sum_n X_n Y_n \sin(\Phi_n - \Psi_n)(\sin M_{nb}\,\Delta\theta_b) &= 0 \\
\sum_n X_n Y_n \cos(\Phi_n - \Psi_n)M_{nb}^2 &> 0 \\
\sum_n X_n Y_n \sin(\Phi_n - \Psi_n)M_{nb} &= 0
\end{aligned}
\tag{4.26}
$$

The above are the most useful evaluations for the formulas and will be used henceforth. It must always be remembered, however, that each of these implies a certain mathematical requirement on x, on y, or on both before such evaluations are valid.

Each of the above formulas actually represents a *set* of relations. Since $\Delta\theta_b$ is arbitrary the first two formulas represent infinite sets for each base frequency. The last two formulas are independent of $\delta\theta_b$ so there can be only one summation of each type for each base frequency.

DISSIPATIVE DEVICES

Dissipative devices may be defined as those and only those wherein the dimensions of the input and output waves are such that the product of x and y at any time gives the instantaneous power at that time. This is indicated by the shaded area in Figure 4.3 where x and y might be the voltage and current in an electrical resistor or the force and velocity in a mechanical damper. If x is a composite wave then for any frequency the time dependent power has the form

$$x_n y_n = X_n \cos(\theta_n + \Phi_n) Y_n \cos(\theta_n + \Psi_n)$$

$$= \tfrac{1}{2} X_n Y_n \cos(\Phi_n - \Psi_n)[1 + \cos 2(\theta_n + \Phi_n)]$$

$$+ \tfrac{1}{2} X_n Y_n \sin(\Phi_n - \Psi_n)[\sin 2(\theta_n + \Phi_n)] \tag{4.27}$$

The first term has both a constant and an oscillatory component. The amplitudes are such that the sum is not negative. The time average of this term is called the *real power* and is given by the constant

$$P_n = \tfrac{1}{2} X_n Y_n \cos(\Phi_n - \Psi_n) \tag{4.28}$$

The second term has the same dimension as the first and also represents a flow of energy. However, there is no constant and hence this term indicates energy which flows into and out of the device. This form of circulating energy is most typical of a class of devices known as *reactive* elements so this form of

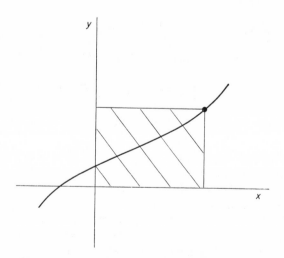

Figure 4.3 Instantaneous power in a dissipative device.

power is called *reactive power*. Its amplitude is given by

$$\mathscr{P}_n = \tfrac{1}{2} X_n Y_n \sin (\Phi_n - \Psi_n) \tag{4.29}$$

Since all components of x and y are in phase if the dissipative element is linear it is impossible for reactive power to exist in a linear dissipative device. It is somewhat astonishing to learn that reactive power can appear in nonlinear dissipative devices. This can be demonstrated quite simply by assuming an input

$$x = \sin \theta + \cos 3\theta \tag{4.30}$$

to a device represented by $y = x^3$. The third harmonic generated by the nonlinearity is in quadrature with the third harmonic component in the input and reactive power appears. This form of reactive power does not represent energy storage such as that which appears in a truly reactive device but instead it indicates the presence of phase relationships between components of x and y which are similar to those in reactive devices and which indicate an oscillatory flow of energy. It should be observed that although certain components of the power may be oscillatory the *total* instantaneous power is always positive in a dissipative device.

The basic canonical relations given by Equations 4.26 contain explicitly the expressions for real and reactive power. A direct substitution gives the power formulas for dissipative devices,

$$\sum_n P_n(1 - \cos M_{nb} \Delta\theta_b) > 0$$
$$\sum_n P_n M_{nb}{}^2 > 0$$
$$\sum_n \mathscr{P}_n \sin M_{nb} \Delta\theta_b = 0$$
$$\sum_n \mathscr{P}_n M_{nb} = 0$$

$$\tag{4.31}$$

By definition the product xy gives the instantaneous power for all dissipative devices and for only such devices. Therefore the above formulas are necessary and sufficient.

The sense of the above inequalities is derived from the assumption that $y(x)$ has no regions of negative slope. However, for dissipative devices the sense of the physical quantities represented by x and y is such that the characteristic curve may lie in only the first and third quadrants. There is no requirement on the slope. Although the slope is generally positive there may be regions of negative slope. Therefore the above inequalities do not necessarily apply to all dissipative devices. Instead, if the input signal is limited entirely to regions of negative slope then the sense of the inequalities should be reversed and the device behaves dynamically as a source rather than a sink.

The formulas given in Equations 4.31 are approximately in the forms in which they were originally presented. The first inequality was reported by Page in 1956 and the second inequality was presented by Pantell in 1958. The last equality was first reported in this country by Manley and Rowe in 1956 although Kudrewicz states that a similar result was formulated by Groszkowski in 1932. The development by Manley and Rowe was based on Fourier integrals and is applicable to only two base frequencies. Their work was generalized by Page in 1957. The original reports of all of these investigators are listed as references for this chapter. From Equations 4.31 a few fundamental theorems may be derived.

Theorem 4.3 *If a dissipative device generates a single subharmonic when driven by a composite wave derived from only one base frequency then it behaves as a sink at this subharmonic frequency.*

Assuming only one base frequency the Page formula can be written as

$$\sum_n P_n(1 - \cos M_n\theta) > 0 \qquad (4.32)$$

Let $\theta = 2\pi$ and the weighting factor for the fundamental and all harmonics is zero. For any subharmonic this factor is necessarily positive. Therefore P_n must be positive and this sense indicates dissipation. It should be observed since the weighting factor is not constant it may not be concluded that *all* subharmonics are necessarily positive if several exist simultaneously. If the subharmonic numbers are known then in special cases the consideration of unique values for θ may allow such conclusions to be established. An example is given in the following corollary.

Corollary 4.3.1 *If the lowest subharmonic in a dissipative device is prime then the device is a sink at this subharmonic frequency.*

For a subharmonic to be prime implies that M_n is the reciprocal of n where n is a prime number. Then the selection of θ by the relation

$$\theta = (n - 1)! \, 2\pi \qquad (4.33)$$

causes the factor $(1 - \cos M_n\theta)$ to be zero for all higher frequencies. Since n is prime it cannot be zero for the subharmonic frequency. The sense of the Page inequality can be preserved only if P_n is positive.

Theorem 4.4 *When a dissipative device is used as a modulator the power available in first order modulation products cannot exceed the power available in the signal or the carrier.*

The carrier with angular velocity ω_c and the signal with angular velocity ω_s must be incommensurable base frequencies. A *modulation product* is a signal whose angular velocity is a linear combination of these base frequencies. Considering a phase displacement in only the signal generates one Page inequality with the form

$$\sum_n P_n (1 - \cos M_{ns} \Delta\theta_s) > 0 \qquad (4.34)$$

For the carrier and harmonics of the carrier $M_{ns} = 0$ and the weighting factor is zero. Therefore the sum may be formed over only the signal, its harmonics, and the modulation products. In normal modulation systems no power is injected or withdrawn at frequencies which are harmonics of the signal since this indicates distortion of the original wave. For first order modulation products $M_{ns} = \pm 1$. Choosing $\Delta\theta_s = 90°$ reduces Equation 4.34 to the form

$$\sum_n P_n > 0 \qquad (4.35)$$

Since it is desired to *extract* power in the modulation products this can be written as

$$|P_m| < |P_s| \qquad (4.36)$$

There is no restriction on M_{nc} so P_m indicates the total power available in all first order modulation products relative not only to the carrier but also to harmonics of the carrier. Normally such higher order products are of no interest and usually they are minimized by auxiliary circuits. Equation 4.36 remains valid for the more common case involving only one or two sidebands. It is also valid whenever M_{ns} is any odd number but again such higher order products are of little practical value.

This theorem can be used advantageously in the design of radio, television, and other communication equipment wherein modulation is performed at relatively high power levels with devices such as vacuum tubes or semiconductors which are intrinsically dissipative. In such applications it must be remembered that the power flow at different frequencies must be assigned appropriate algebraic signs depending on whether it is being absorbed or being made available at the two terminals of a purely dissipative element. It is not valid, for example, to include grid circuit losses when applying the theorem to the dissipative element consisting of the plate to cathode circuit in a vacuum tube.

Theorem 4.5 *The efficiency of harmonic generation by a dissipative device cannot exceed the reciprocal of the square of the harmonic number.*

This is established by assuming that power is delivered at only one harmonic frequency since this must be the most efficient situation. Applying the Pantell inequality to this case gives

$$P_1 + P_n n^2 > 0 \qquad (4.37)$$

Since P_n must be negative this is equivalent to

$$\frac{|P_n|}{P_1} < \frac{1}{n^2} \tag{4.38}$$

and this establishes the theorem. This result is completely general and applies to all dissipative elements. Since common electronic tools such as vacuum tubes and many semiconductor devices are included in this category designers using such dissipative elements for frequency multiplication must recognize this fundamental limit on efficiency. A few examples will be given to illustrate this point.

Perhaps the simplest nonlinear element which has easily expressible and appropriate characteristics is the electrical resistor wherein $v = i^3$. This is an odd function with positive slope everywhere except at the origin. When a current $i_1 = \cos \theta$ passes through this resistor the voltage across it can be expressed as

$$v = i^3 = \tfrac{3}{4} \cos \theta + \tfrac{1}{4} \cos 3\theta \tag{4.39}$$

The resistor can be considered as a possible source of power at the third harmonic. To use this power the network shown in Figure 4.4 may be employed. It is assumed that the filters present zero impedance at the indicated frequencies and infinite impedance at all others. Thus the two loops in the network operate at different frequencies and there is no connection between the generator and the load.

Assuming in the first case that the load is purely resistive then the current i_3 will be in phase with the third harmonic voltage and the total current through the nonlinear resistor will be

$$i = i_1 - i_3 = \cos \theta - I_3 \cos 3\theta \tag{4.40}$$

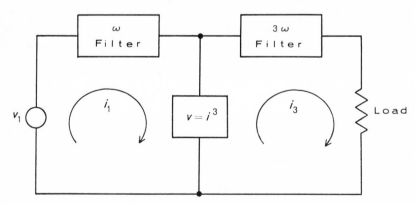

Figure 4.4 Harmonic generation with a nonlinear resistor.

This produces a voltage of the form

$$v = i^3 = \tfrac{3}{4}(1 - I_3 + 2I_3{}^2) \cos \theta$$
$$+ \tfrac{1}{4}(1 - 6I_3 - 3I_3{}^3) \cos 3\theta$$
$$+ \text{ irrelevant terms} \qquad (4.41)$$

The amounts of power delivered by the generator, absorbed by the nonlinear resistor, and dissipated in the load may be expressed respectively as

$$P_g = \langle i_1 v \rangle \quad\ = \tfrac{3}{8}(1 - I_3 + 2I_3{}^2)$$
$$P_r = \langle (i_1 - i_3)v \rangle = \tfrac{1}{8}(3 - 4I_3 + 12I_3{}^2 + 3I_3{}^4) \qquad (4.42)$$
$$P_l = \langle i_3 v \rangle \quad\ = \tfrac{1}{8}(I_3 - 6I_3{}^2 - 3I_3{}^4)$$

Since the value of the load resistance is arbitrary it may be selected so as to maximize the output power. This is done by setting the derivative of P_l with respect to I_3 equal to zero. This provides the relation

$$12I_3{}^3 + 12I_3 - 1 = 0 \qquad (4.43)$$

which is solved approximately by $I_3 = 0.083$ ampere. This current is drawn by a resistance of 1.5 ohm. This is the load which is "matched" to the nonlinear resistor in this operating mode. Substituting the selected value for I_3 gives

$$P_g = 0.349 \text{ watt}$$

$$P_r = 0.344 \text{ watt} \qquad (4.44)$$

$$P_l = 0.005 \text{ watt}$$

Although the load was selected so as to maximize the output power the device is highly inefficient. Theorem 4.5 permits a maximum efficiency of 11.1% but in reality it is around 1.4%.

When the load is disconnected the voltage at the fundamental frequency across the nonlinear resistor is 0.75 volt and the resistor dissipates 0.375 watt as heat. When the load is connected this voltage drops to around 0.70 volt and the resistor draws only 0.349 watt. Of this total power 0.344 watt is dissipated as heat and 0.005 watt is transformed into power at the third harmonic and is delivered to the load. Although the filters prevent a direct connection between the generator and the load the nonlinear resistor acts as a coupling device and accomplishes a transfer of power.

An interesting thing happens if the load impedance contains reactance as well as resistance. The third harmonic current is no longer in phase with the voltage so the total current through the nonlinear resistor must be written in the general form

$$i = \cos \theta - I_3 \cos (3\theta + \Phi) \qquad (4.45)$$

The voltage across the resistor becomes

$$v = i^3 = (\tfrac{3}{4} + \tfrac{3}{2}I_3{}^2) \cos \theta - \tfrac{3}{4}I_3 \cos (\theta + \Phi) \qquad (4.46)$$
$$+ \tfrac{1}{4} \cos 3\theta - (\tfrac{3}{2}I_3 + \tfrac{3}{4}I_3{}^3) \cos (3\theta + \Phi)$$
$$+ \text{irrelevant terms}$$

It is observed that both real and reactive power exist at both the fundamental and third harmonic frequencies. Calculating and combining the various components give

$$P_1 = \tfrac{3}{8} + \tfrac{3}{4}I_3{}^2 - \tfrac{3}{8}I_3 \cos \Phi$$
$$P_3 = \tfrac{3}{4}I_3{}^2 + \tfrac{3}{8}I_3{}^4 - \tfrac{1}{8} \cos \Phi$$
$$\mathscr{P}_1 = -\tfrac{3}{8}I_3 \sin \Phi \qquad (4.47)$$
$$\mathscr{P}_3 = \tfrac{1}{8} \sin \Phi$$

In the above expressions reactive power with the phase relationship found in inductors is considered to be positive and that found in capacitors is negative.

In accordance with the phase relationships stated in Equations 4.45 and 4.46 the load must be capacitive if Φ is positive. As an example, if Φ is positive then $\mathscr{P}_1$ is negative and $\mathscr{P}_3$ is positive. The resistor draws negative inductive energy, hence capacitive energy, from the source and delivers positive inductive energy to the load. Here it appears as capacitive energy since the sense of the current is reversed. Again the nonlinear resistor appears to couple the load to the generator although in this more complex case it also accomplishes a transfer of reactive power.

The load impedance determines I_3 and Φ to a limited extent. Since the load is a linear circuit element operating at only one frequency conventional analysis may be employed. Applying appropriate trigonometric identities to Equation 4.46 permits the third harmonic current to be expressed as

$$i_3 = I_3 \cos (3\theta + \Phi)$$
$$v_3 = \tfrac{1}{4} \sin \Phi \sin (3\theta + \Phi) \qquad (4.48)$$
$$+ (\tfrac{1}{4} \cos \Phi - \tfrac{3}{2}I_3 - \tfrac{3}{4}I_3{}^3) \cos (3\theta + \Phi)$$

Assuming that the load impedance has components designated as $R + jX$ and using $\sin (3\theta + \Phi)$ as the phase reference permit voltage relationships to be expressed as

$$\tfrac{1}{4} \sin \Phi + j(\tfrac{1}{4} \cos \Phi - \tfrac{3}{2}I_3 - \tfrac{3}{4}I_3{}^3) = jI_3(R + jX) \qquad (4.49)$$

Equating real and imaginary terms gives

$$\sin \Phi = -4I_3 X$$
$$\cos \Phi = 4I_3 R + 6I_3 + 3I_3{}^3 \qquad (4.50)$$

These may be solved simultaneously for I_3 and Φ as functions of R and X if desired.

To complete this example with a numerical result it will be assumed that $I_3 = 0.1$ ampere and $R = 0.5$ ohm. This gives

$$\sin \Phi = 0.558$$
$$\cos \Phi = 0.830 \tag{4.51}$$

and the required amount of reactance is $X = -0.139$ ohm. This is capacitive since Φ is considered to be positive. These numerical values give the following results for the real and reactive power flow in the nonlinear resistor

$$P_1 = +0.322 \text{ watt}$$
$$P_3 = -0.012 \text{ watt}$$
$$\mathscr{P}_1 = -0.021 \text{ watt} \tag{4.52}$$
$$\mathscr{P}_3 = +0.007 \text{ watt}$$

It is immediately apparent that all of the basic power formulas given in Equations 4.31 are satisfied. Of the total input power 0.310 watt is dissipated as heat in the nonlinear resistor and 0.012 watt is delivered to the load where it is dissipated in the resistive component of the load. The nonlinear resistor also draws 0.021 watt of capacitive reactive power at the fundamental frequency which reappears as 0.007 watt of capacitive reactive power circulating at the third harmonic frequency in the reactive component of the load. This is quite interesting performance for a simple resistor.

The previous examples are remarkably inefficient. Theorem 4.5 gives an upper bound which is much higher. The characteristic which produces harmonic power at maximum efficiency is shown in Figure 4.5. The nonlinear resistor is an ideal limiter with the characteristic shown. This, of course, is not a realizable characteristic and the following discussion is purely hypothetical. Also indicated in the figure are various components of voltage and current in the proper phase. The limiter acts as a source for the harmonic current and the relative phase between voltage and current is reversed.

It is impossible for the harmonic current to cause the limiter to function at the harmonic frequency so the sum of the currents must have the sense of the fundamental component. Therefore the harmonic current must have a slope at the origin whose absolute value is no greater than that of the fundamental component. In general the slopes are given by

$$\frac{di_1}{dt} = \omega I_1 \cos \omega t, \qquad \frac{di_n}{dt} = -n\omega I_n \cos n\omega t \tag{4.53}$$

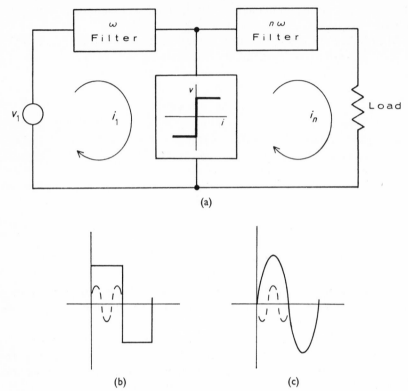

Figure 4.5 Optimum harmonic generation with a dissipative device. (a) The non-linear characteristic, (b) the fundamental and third harmonic voltages, and (c) the currents which flow.

So at the origin the maximum value for I_n is

$$|I_n| = \frac{I_1}{n} \tag{4.54}$$

A square wave was analyzed in the previous chapter and according to Equation 3.59 each harmonic amplitude is

$$V_n = \frac{V_1}{n} \tag{4.55}$$

This produces optimum power generation where

$$|P_n| = \frac{P_1}{n^2} \tag{4.56}$$

Any effort to improve this performance by the use of any other resistor with a positive, monotonic characteristic must fail.

REACTIVE DEVICES

Reactive devices may be defined as those and only those wherein the dimensions of the input and output waves are such that the product of x and y has the dimension of *energy*. Such a product is related to but not equal to the energy stored in the device. As shown in Figure 4.6 the stored energy is the area under the characteristic curve. This should be compared to the situation for dissipative devices shown in Figure 4.3. Typical reactive elements are the electrical inductor or capacitor and mechanical devices with inertial or elastic properties. For these devices the instantaneous energy is expressed in general by

$$\int_0^x y \, dx = \int_0^t y\dot{x} \, dt \tag{4.57}$$

where $\dot{x}$ is the time derivative of x. The integrand is the instantaneous power and for any frequency it has the form

$$\begin{aligned} y_n\dot{x}_n = \tfrac{1}{2}\omega_n X_n Y_n \sin(\Phi_n - \Psi_n)[1 + \sin 2(\theta_n - \Phi_n)] \\ + \tfrac{1}{2}\omega_n X_n Y_n \cos(\Phi_n - \Psi_n)[\cos 2(\theta_n - \Phi_n)] \end{aligned} \tag{4.58}$$

Comparison with Equation 4.27 shows that for reactive devices

$$\begin{aligned} P_n &= \tfrac{1}{2}\omega_n X_n Y_n \sin(\Phi_n - \Psi_n) \\ \mathscr{P}_n &= \tfrac{1}{2}\omega_n X_n Y_n \cos(\Phi_n - \Psi_n) \end{aligned} \tag{4.59}$$

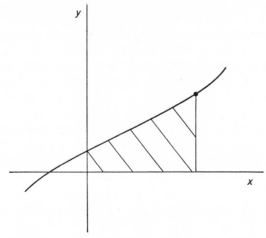

Figure 4.6 The instantaneous energy in a reactive device.

It should be observed that the trigonometric functions are reversed and that the power is proportional to the frequency. Substitution of these relations into the basic canonical equations gives the power formulas for reactive devices

$$\sum_n \frac{\mathscr{P}_n}{\omega_n}(1 - \cos M_{nb}\Delta\theta_b) > 0$$

$$\sum_n \frac{\mathscr{P}_n}{\omega_n} M_{nb}^2 > 0$$

$$\sum_n \frac{P_n}{\omega_n}(\sin M_{nb}\Delta\theta_b) = 0 \qquad (4.60)$$

$$\sum_n \frac{P_n}{\omega_n} M_{nb} = 0$$

The expression for reactive power in Equation 4.59 is symmetrical with respect to the amplitudes and phase angles so there is no clear method for assigning polarity to the various types of reactive power. Correlating physical and mathematical variables so that $y\,dx$ corresponds to an element of stored energy in each case shows that the electrical variables of charge and flux correspond to x and that voltage and current correspond to y. When the phase relation between voltage and current is calculated in each case it is found that the reactive power is positive in each case. This shows that *any* form of reactive power can be considered to be positive at an appropriate time in the above equations.

This situation an be reconciled by the following procedure. If when examined at its two external terminals the device is described by a voltage versus charge characteristic curve then the convention of signs must be such that capacitive reactive power is positive and the inequalities of Equation 4.60 will be valid. A realizable capacitance has inductance in its leads and along its surface. This has the effect of superimposing negative inductive reactive power at all of the frequencies present but hopefully this will not disturb Equation 4.60. It follows directly that if the device is described by a flux versus current characteristic then inductive reactive power is considered to be positive. Application of the inequalities to mechanical vibrations must follow corresponding rules. Due to these problems of polarity the above inequalities are rarely used.

The same is not true of the equalities, however. The latter, which is due to Manley and Rowe, is undoubtedly the most widely used of all the power formulas. The popularity stems mostly from the development of highly efficient nonlinear capacitors using semiconductor materials. Since the normal mode of a reactive device is to store energy rather than to dissipate it the presence of real power in such a device offers an excellent means for power conversion. Several important theorems can be established rather easily although a few are somewhat puzzling.

Theorem 4.6 *When an ideal nonlinear reactive device is used as a frequency multiplier the efficiency is limited exclusively by the external circuit.*

For this special case where there is only one base frequency and power is extracted at only one harmonic frequency the Manley-Rowe equations reduce to the single expression

$$\frac{P_1}{\omega_1} - \frac{P_n n}{n\omega_1} = 0 \tag{4.61}$$

which reduces immediately to

$$P_n = P_1 \tag{4.62}$$

and the theorem is established. It is interesting to observe that whenever there is a single base frequency the Manley-Rowe equations reduce to the expression of conservation of power.

Theorem 4.7 *When a reactive device is used to translate frequencies the absolute value of the power absorbed or delivered at each frequency is proportional to that frequency.*

By frequency translation is meant the process whereby power is injected at two incommensurable frequencies and extracted at a third which is either the sum or difference but not a harmonic. Although other frequencies may be generated it is assumed that the load presents either an open circuit, a short circuit, or a pure reactance at all other frequencies. Since there are two base frequencies there are the two relations

$$
\begin{aligned}
\frac{P_1}{\omega_1} + \frac{P_\pm}{\omega_1 \pm \omega_2} &= 0 \\
\frac{P_2}{\omega_2} - \frac{P_\pm}{\omega_1 \pm \omega_2} &= 0
\end{aligned}
\tag{4.63}
$$

where the negative signs must be used appropriately with difference frequencies and $P_\pm$ symbolizes the output power. These may be solved simultaneously to give

$$\frac{P_1}{\omega_1} = \pm \frac{P_2}{\omega_2} = - \frac{P_\pm}{\omega_1 \pm \omega_2} \tag{4.64}$$

which establishes the theorem. This result is at first puzzling because it states that it is impossible to force a reactive device to accept or deliver power in any but the above ratios. There is no latitude. This is illustrated in Figure 4.7. The result is more acceptable, however, when it is realized that the theorem applies only to real power in a purely reactive device. In all realizable networks there are dissipative elements in the generators and in the reactive device so departures from the above requirements are experienced.

Theorem 4.8   *When a reactive device is used as a modulator to produce an upper sideband the system is stable and provides a power gain.*

It is assumed that the signal has the angular velocity ω_1 and that the carrier has the angular velocity ω_2. The upper sideband then has the angular velocity $\omega_1 + \omega_2$ and the power formulas are

$$\frac{P_1}{\omega_1} + \frac{P_+}{\omega_1 + \omega_2} = 0$$

$$\frac{P_2}{\omega_2} + \frac{P_+}{\omega_1 + \omega_2} = 0 \tag{4.65}$$

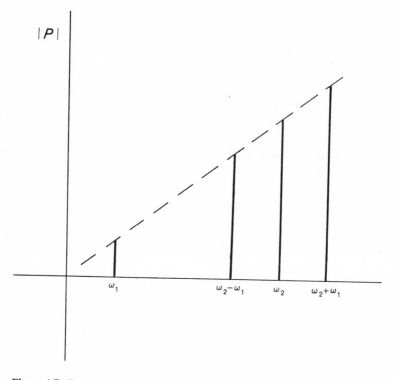

Figure 4.7   Power in a reactive device used to translate frequencies.

which can be solved simultaneously to provide the relations

$$\frac{P_1}{\omega_1} = \frac{P_2}{\omega_2} = -\frac{P_+}{\omega_1 + \omega_2} \tag{4.66}$$

In this application P_1 and P_2 must be positive and P_+ must be negative. This power flow is consistent with Equation 4.66 so the performance of the device is compatible with its requirements and the system is stable. The power gain is given by

$$G_m = \frac{|P_+|}{P_1} = \frac{\omega_1 + \omega_2}{\omega_1} > 1 \tag{4.67}$$

If the carrier frequency is high compared to the signal frequency this system can provide a very high power gain.

Corollary 4.8.1 *Demodulation of an upper sideband by a reactive device is a potentially unstable process and provides a power loss.*

If a reactive device is used to demodulate the upper sideband and restore the signal then the situation is much different. Equations 4.65 and 4.66 still apply but now according to the application P_2 and P_+ should be positive and P_1 should be negative, but according to Equation 4.66 P_+ is positive and both P_2 and P_1 are negative. This is a potentially unstable system since the reactive device delivers power into the local oscillator at angular velocity ω_2 instead of accepting power from it. This system is stabilized by introducing positive resistance into the local oscillator circuit.

Assuming stability in the local oscillator circuit the gain of the demodulator is given by

$$G_d = \frac{|P_1|}{P_+} = \frac{\omega_1}{\omega_1 + \omega_2} < 1 \tag{4.68}$$

This is the reciprocal of the gain of the modulator so the use of reactive devices for both functions provides no net gain. As an illustration of this principle demodulation in magnetic amplifiers is frequently performed with diode rectifiers.

Theorem 4.9 *When a reactive device is used as a modulator to produce a lower sideband the system is potentially unstable.*

In this case the sideband has angular velocity $\omega_2 - \omega_1$ and the power formulas are

$$\frac{P_1}{\omega_1} - \frac{P_-}{\omega_2 - \omega_1} = 0$$

$$\frac{P_2}{\omega_2} + \frac{P_-}{\omega_2 - \omega_1} = 0 \tag{4.69}$$

which can be solved simultaneously to provide the relations

$$-\frac{P_1}{\omega_1} = \frac{P_2}{\omega_2} = -\frac{P_-}{\omega_2 - \omega_1} \tag{4.70}$$

Since ω_2 is larger than ω_1 these relations are satisfied when P_1 and P_- are negative and P_2 is positive. This indicates that the local oscillator supplies power at angular velocity ω_2 to the modulator as it should. Also, the modulator supplies output power at the lower sideband frequency as it should. However, the reactive device can also supply power at the signal frequency. This permits an unstable system to be formed. After a signal has been injected into the system and a mode of operation established then the removal of the signal may not cause the output to vanish. If the power which the reactive device introduces to the input circuit exceeds the losses in that circuit then the modulator will continue to supply output power. The frequency will probably be close to the original but the exact value depends on external circuit elements. This type of modulator is stabilized by the addition of positive resistance to the input circuit.

Due to the above possibility of unstable operation it is meaningless to calculate the gain in this type of modulator. A large amount of gain can be achieved by operating near the point of instability if this is an acceptable mode.

Demodulation of the lower sideband also involves Equations 4.69 and 4.70 with the same polarity for P_1, P_2, and P_-. The situation is identical and the same stability problem exists.

There is a type of signal used in telephone and radio communication wherein both sidebands are transmitted and the carrier is suppressed. This signal is normally generated at low levels in balanced circuits using dissipative elements. However, to indicate further application of the power formulas the generation of such signals with reactive devices will be discussed. Again considering a signal with angular velocity ω_1 and a carrier with angular velocity ω_2 as incommensurable base frequencies there are the two power formulas

$$\frac{P_1}{\omega_1} - \frac{P_-}{\omega_2 - \omega_1} + \frac{P_+}{\omega_2 + \omega_1} = 0$$

$$\frac{P_2}{\omega_2} + \frac{P_-}{\omega_2 - \omega_1} + \frac{P_+}{\omega_2 + \omega_1} = 0 \tag{4.71}$$

Adding and subtracting these equations provide the relations

$$\frac{P_1}{\omega_1} + \frac{P_2}{\omega_2} = \frac{-2P_+}{\omega_2 + \omega_1}$$

$$\frac{P_1}{\omega_1} - \frac{P_2}{\omega_2} = \frac{2P_-}{\omega_2 - \omega_1} \tag{4.72}$$

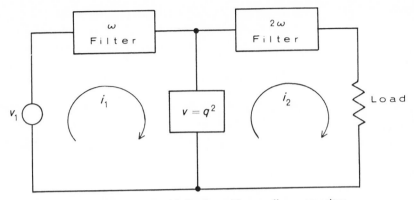

Figure 4.8 Frequency multiplication with a nonlinear capacitor.

The first equation shows that upper sideband power can be extracted for all levels of P_1 and P_2. The second equation shows that lower sideband power can be extracted only if there is adequate carrier power as given by

$$\frac{P_2}{\omega_2} > \frac{P_1}{\omega_1} \tag{4.73}$$

Adding and subtracting Equations 4.72 give

$$\frac{P_1}{\omega_1} = \frac{P_-}{\omega_2 - \omega_1} - \frac{P_+}{\omega_2 + \omega_1}$$

$$\frac{P_2}{\omega_2} = - \frac{P_-}{\omega_2 - \omega_1} - \frac{P_+}{\omega_2 + \omega_1} \tag{4.74}$$

Assuming that $\omega_2 \gg \omega_1$ these can be replaced with the approximations

$$\left(\frac{\omega_2}{\omega_1}\right) P_1 = P_- - P_+$$

$$P_1 = -P_- - P_+ \tag{4.75}$$

The first equation shows that the upper sideband contains more power and the excess is proportional to the signal strength. The second equation shows that the total sideband power is proportional to the carrier strength. These peculiar results must again be reconciled by the realization that they apply only to purely reactive devices.

The previous results will be illustrated by an example wherein frequency multiplication is performed by a nonlinear capacitor with a quadratic characteristic. The network is shown in Figure 4.8. The total current through the capacitor may be written as

$$i = i_1 - i_2 = I_1 \cos \omega t - I_2 \cos (2\omega t + \Phi) \tag{4.76}$$

where I_2 is positive. The charge at any time is given by

$$q = \frac{I_1}{\omega} \sin \omega t - \frac{I_2}{2\omega} \sin (2\omega t + \Phi) \qquad (4.77)$$

The voltage across the capacitor contains the components expressed by

$$v = -\frac{I_1 I_2}{2\omega^2} \cos (\omega t + \Phi) - \frac{I_1{}^2}{2\omega^2} \cos 2\omega t + \text{irrelevant terms} \qquad (4.78)$$

The power flow at the fundamental and harmonic frequencies therefore is given by the relations

$$P_1 = -\frac{I_1{}^2 I_2}{4\omega^2} \cos \Phi$$

$$P_2 = +\frac{I_1{}^2 I_2}{4\omega^2} \cos \Phi \qquad (4.79)$$

Proper sense is provided if Φ lies between $90°$ and $270°$. It is not surprising that the two expressions are numerically identical since power must be conserved. As there is only one base frequency Equations 4.60 require nothing more. The inequalities do not apply since the nonlinear characteristic does not have positive or zero slope everywhere.

An equivalent circuit for a frequency multiplier using a nonlinear reactor is shown in Figure 4.9. The impedances Z_1 and Z_m represent the effects of filters which pass only the fundamental and harmonic frequencies. The resistance r is introduced to account for losses in the reactor. These will be resistive and magnetic circuit losses in an inductor and dielectric losses in a

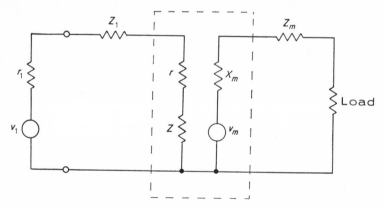

Figure 4.9 Equivalent circuit for a frequency multiplier using a nonlinear reactor.

capacitor. The impedance Z is selected to provide the proper value for the current i_1. The contrived source v_m represents the harmonic generation by the nonlinear properties of the reactor. The effective internal impedance of the device at the harmonic frequency is represented by X_m. The exact values for these network elements can be found either experimentally or by a detailed analysis using an appropriate mathematical expression for the nonlinear characteristic.

Before this discussion of reactive devices is concluded a few examples will be given wherein power formulas of the Manley-Rowe type appear from quite different considerations. Three examples are given and all of the original reports are listed as references for this chapter. The first is an alternate derivation due to Salzberg.

Since in a reactive device the power is proportional to the frequency it can be expressed as

$$P = fW \tag{4.80}$$

where f is the frequency and W is the energy per cycle. For a process which combines frequencies according to the relation

$$f_3 = M_1 f_1 + M_2 f_2 \tag{4.81}$$

the expression of conservation of power becomes

$$f_1 W_1 + f_2 W_2 + f_3 W_3 = f_1(W_1 + M_1 W_3) + f_2(W_2 + M_2 W_3) = 0 \tag{4.82}$$

Since f_1 and f_2 are independent base frequencies the only permissible solutions are

$$W_1 + M_1 W_3 = 0$$
$$W_2 + M_2 W_3 = 0 \tag{4.83}$$

Substituting the original relations and introducing the angular velocity give the result

$$\frac{P_1}{\omega_1} + \frac{P_3 M_1}{\omega_3} = 0$$
$$\frac{P_2}{\omega_2} + \frac{P_3 M_2}{\omega_3} = 0 \tag{4.84}$$

These are the Manley-Rowe equations for such a process.

Weiss has shown the initially startling fact that when a group of atoms are excited from an external power source the imposition of quantization of energy levels leads to equations of the Manley-Rowe type. It is assumed that in a certain type of atom the electrons can be in three possible states which are designated a, b, and c. When passing in transition from one state to another energy is transmitted or absorbed and if radiation is involved then the frequency of the radiation is proportional to the difference in energy levels between the two states. Therefore in general the transitional power can be expressed as

$$P = Nhf \tag{4.85}$$

where N is the number of transitions per second and h is known as Planck's constant which relates the frequency and the energy at each transition.

Under steady state conditions to conserve energy it is sufficient that the total number of electrons in each state remains constant. This is equivalent to the requirements

$$N_{ab} + N_{bc} = 0$$
$$N_{bc} + N_{ca} = 0 \tag{4.86}$$

where the subscripts indicate the directions and energy levels where the transitions occur. Substituting the original relation gives the result

$$\frac{P_1}{\omega_1} + \frac{P_2}{\omega_2} = 0$$
$$\frac{P_2}{\omega_2} + \frac{P_3}{\omega_3} = 0 \tag{4.87}$$

These are the Manley-Rowe equations which apply to frequency translation in a reactive device. In accord with the above derivation they also apply to the atomic device known as a *laser*.

A more unusual derivation of the Manley-Rowe type of power formula has been given by Pierce. His method considers the reflection of a plane wave of electromagnetic energy from a plane surface moving with velocity v relative to the observer. Equating frequencies at the reflecting surface gives the relation

$$\frac{\omega_2}{\omega_1} = \frac{c + v}{c - v} \tag{4.88}$$

where c is the velocity of the wave. It is a consequence of Maxwell's equations that the time rate of change of momentum of such a plane wave is the power divided by the velocity so a statement of conservation of momentum provides the equation

$$F = \frac{P_2 + P_1}{c} \tag{4.89}$$

where F is the force which the reflecting surface exerts on the wave. Conservation of energy gives the additional requirement

$$F = \frac{P_2 - P_1}{v} \tag{4.90}$$

Combining the above gives the relation

$$\frac{P_2}{P_1} = \frac{c + v}{c - v} \tag{4.91}$$

Comparing this with the original relation between frequencies gives the result

$$\frac{P_1}{\omega_1} = \frac{P_2}{\omega_2} \tag{4.92}$$

which is a Manley-Rowe form for frequency translation.

The repeated occurrence of power formulas of the Manley-Rowe type in such diverse situations has aroused the speculation that there possibly are more general requirements which transcend all of the examples given. Such expressions have never been found. It is perhaps more rational to assume the attitude that whenever processes involve conservation of energy and situations wherein power is proportional to frequency then the power formulas should be mutually consistent. Such appears to be the case.

HYSTERESIS

The previous results have depended partly on the assumption that the nonlinear characteristic is single valued. This was specifically assumed in the interpretation of Equation 4.25 to provide the evaluation of the last equality in Equations 4.26. This led to the Manley-Rowe equations which have a particularly important position in describing the power flow in reactive devices.

Unfortunately it is true that reactive devices frequently have a multiple-valued nonlinear characteristic. This common phenomenon is known as hysteresis and is illustrated in Figure 4.10. When traversing the curve from A to B there is energy stored in the device equal to the area under the curve. Sometimes the energy storage causes a physical change in the device such that when an attempt is made to recover this energy the return path from B to A does not coincide with the original path. The device fails to deliver all of the energy which was placed in it. The retained energy equals the shaded area in Figure 4.10.

If an average of the two curves is drawn from A to B it is observed that the departure is negative from A to B and positive from B to A. This shows the presence of a wave in quadrature with the original input wave. In a reactive device this indicates real power and in a dissipative device this indicates the presence of reactive power.

When the input signal contains composite waves derived from several base frequencies the path becomes very complex and the corresponding mathematical analysis must be generalized. The integral in Equation 4.25 which

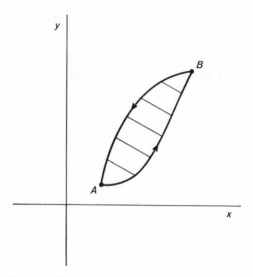

Figure 4.10 Performance of a device with hysteresis.

provides the null condition for the Manley-Rowe equations must now be evaluated by the expressions

$$- \lim_{T \to \infty} \frac{1}{\omega_b T} \int_0^{t=T} y \, dx = \frac{1}{\omega_b} \sum_n H_n = \frac{H}{\omega_b} \qquad (4.93)$$

where H_n is the power loss at each frequency due to hysteresis and H is the total power loss due to hysteresis. The power formulas assume the general form

$$\sum_n \frac{P_n}{\omega_n} M_{nb} = \frac{H}{\omega_b} \qquad (4.94)$$

If there is predominant excitation at one base frequency it is sometimes acceptable to ignore losses at all other frequencies and combine H_n with P_n to restore the null condition. In this case the hysteresis loss subtracts directly from the generator power at that frequency.

GENERALIZATIONS

The first generalization of the power formulas which will be discussed involves their application to devices whose properties vary in time. The evaluation of the canonical equations which leads to the most useful forms for the power formulas requires only that the nonlinear characteristic be single valued and have everywhere a positive or zero slope. There is no further requirement. It is apparent therefore that the power formulas remain valid when the nonlinear characteristic varies in time provided that the above requirements continue to be met.

A new problem arises, however, when the variation is periodic in time. Power of various forms is generated at new frequencies and the question arises as to how this influences the power formulas. The situation will be illustrated by an example. It is assumed that the capacitance of a capacitor is varied sinusoidally about a mean value and that the voltage across it also varies sinusoidally at another frequency. The charge can then be expressed by

$$q(t) = vc = V \sin \omega t (C_0 + C \sin \omega_c t) \tag{4.95}$$

and the current then has the form

$$i(t) = \omega V C_0 \cos \omega t$$
$$- \frac{VC}{2} [(\omega - \omega_c) \sin (\omega - \omega_c)t - (\omega + \omega_c) \sin (\omega + \omega_c)t] \tag{4.96}$$

Comparison with the voltage shows that reactive power exists at angular velocity ω and this is independent of ω_c. This is not astonishing. Since this is a linear model there can be no real power which is independent of ω_c. However, by making $\omega_c = 2\omega$ there is real power which is dependent on ω_c and this may be a bit astonishing. The negative sign indicates that this power is supplied by the capacitor to the external circuit. This power is derived from mechanical sources since mechanical energy is involved in changing the capacitance of a charged capacitor. This means that although the time-varying capacitor conserves energy in the sense that mechanical energy is converted to electrical energy without dissipation it does not appear to conserve energy when viewed strictly from the electrical network variables. The power formulas were derived from a consideration of a closed set of such variables and in particular the Manley-Rowe formulation implies conservation of energy. These considerations are reconciled by the stipulation that all of the power formulas continue to apply to time-varying reactive devices under the restriction that the periodic variation in time shall not be considered as an independent base frequency and that power at all frequencies which is transduced from such variation in time shall be excluded from the sums.

It was previously stated that conservation of energy is a necessary condition for the Manley-Rowe equations. It may be more precise and somewhat more satisfying to state that the Manley-Rowe equations apply also to nonconservative devices but to only that portion of the total activity which is in itself conservative.

There is an interesting and frequently used application of the above principles. Sometimes a nonlinear device is intentionally driven through portions of its nonlinear characteristic to provide a periodic variation in its properties. This situation can be analyzed in two ways. The completely general way is to apply the power formulas directly and to include a summation for each base frequency. The second method is to exclude the base frequency which provides the periodic variation in the properties of the device and to substitute an appropriate variation in time of the nonlinear characteristic. This second method provides information about only those signals which are not dependent on the variation of parameters in time and hence offers some simplification. A power gain can be achieved by the above technique and the device is frequently called a *parametric amplifier.*

It follows by analogy that a corresponding situation exists with a dissipative device. It is particularly interesting to observe that a periodic variation of its characteristic in time produces the equivalent of reactive power and that this is accomplished without the extraction of power from the mechanical device providing the periodic variation in the nonlinear characteristic.

The next generalization of the power formulas will be the interesting, informative, but possibly impractical consideration of devices which are neither dissipative nor reactive. Elementary electrical and mechanical devices appear always to fall into one of those two categories but more complex active devices, for example those used in analog computation, can provide higher orders of integration and differentiation. It may also be possible that mankind will invent simple devices in the future which involve such higher orders. Their power formulas can be derived by a simple extension.

In a dissipative device the instantaneous power is given by $y_n x_n$, the real power involves $\cos(\Phi_n - \Psi_n)$, and the reactive power depends on $\sin(\Phi_n - \Psi_n)$. In a reactive device the instantaneous power is expressed by $y_n \dot{x}_n$, the real power depends on $\sin(\Phi_n - \Psi_n)$, and the reactive power involves $\cos(\Phi_n - \Psi_n)$. Continuing this process it is observed that if the instantaneous power should be given by $y_n \ddot{x}_n$ then the real power would depend on $-\cos(\Phi_n - \Psi_n)$ and the reactive power would involve $-\sin(\Phi_n - \Psi_n)$. Successively higher time derivatives continue to shift the trigonometric functions, the function expressing real power rotating counterclockwise and the function involving reactive power rotating clockwise. In addition each differentiation adds ω_n as a factor. Combining these effects and inserting the result in the last equality in Equations 4.26 provide the generalized Manley-Rowe equations

$$\sum_n \frac{P_n}{\omega_n^k} M_{nb} = 0 \qquad \text{if } k \text{ is odd}$$

$$\sum_n \frac{\mathscr{P}_n}{\omega_n^k} M_{nb} = 0 \qquad \text{if } k \text{ is even}$$

(4.97)

The above procedure obviously applies to all of the basic power formulas. The Manley-Rowe form was selected only because of its simplicity.

It is a bit difficult to imagine a device wherein power storage depends on its time derivative and power depends on integration. If such a device can be realized then the above procedure shows that k is negative and the factor ω_n^k appears in a multiplicative fashion.

The last generalization of the power formulas to be discussed involves their extension to networks which contain more than one element and elements of more than one kind. The Manley-Rowe forms will again be used, not only for simplicity but also for feasibility. Considering a network of k purely dissipative elements there will be bk relations of the form

$$\sum_n \mathscr{P}'_{nk} M_{nb} = 0 \tag{4.98}$$

where the reactive power in each case is internal to each of the elements. Since each sum is independently zero the sum of the sums is also zero. This gives the relation

$$\sum_k \sum_n \mathscr{P}'_{nk} M_{nb} = 0 \tag{4.99}$$

Extending the scope of the summations over n to include all frequencies present in all elements and invoking conservation of reactive power at each frequency provide the simplified form

$$\sum_n \left(\sum_k \mathscr{P}'_{nk} \right) M_{nb} = \sum_n \mathscr{P}_n M_{nb} = 0 \tag{4.100}$$

where $\mathscr{P}_n$ applies to the network terminals. In the above reduction it is necessary to recognize the two types of reactive power and to assign opposite polarity although the choice is arbitrary. Following the same procedure for a network of reactive elements it is found that

$$\sum_n \left(\sum_k P'_{nk} \right) \frac{M_{nb}}{\omega_n} = \sum_n \frac{P_n}{\omega_n} M_{nb} = 0 \tag{4.101}$$

over the network regardless of the types of reactive elements which are contained. It is required, of course, that the nonlinear characteristics of all elements are single valued.

If the network contains a mixture of both types of devices then it is possible first to combine all elements of each type in each branch and write sums of the above forms for each group of elements. Then as a mathematical gesture it is possible to combine the sums into expressions of the form

$$\sum_n \left(\mathscr{P}_n + \frac{P_n}{\omega_n} \right) M_{nb} = 0 \tag{4.102}$$

but these are not generalized to the branch terminals so the extension cannot continue. Real power in the dissipative elements and reactive power in the reactive devices provide additional terms which may destroy the nullity of the equalities. The real and reactive power in the above sums is limited to that present in each of the two groups of similar elements so the expressions have very restricted theoretical interest.

REFERENCES

1. P. Penfield, *Frequency-Power Formulas*, Wiley, New York, 1960.
2. C. H. Page, Frequency Conversion with Positive Nonlinear Resistors, *Journal of Research of the NBS*, Volume 56, 1956, pp. 179–182.
3. R. H. Pantell, General Power Relationships for Positive and Negative Nonlinear Resistive Elements, *Proceedings of the IRE*, Volume 46, 1958, pp. 1910–1913.
4. C. H. Page, Harmonic Generation with Ideal Rectifiers, *Proceedings of the IRE*, Volume 46, 1958, pp. 1738–1740.
5. J. M. Manley and H. E. Rowe, Some General Properties of Nonlinear Elements—Part I. General Energy Relations, *Proceedings of the IRE*, Volume 44, 1956, pp. 904–913.
6. J. Kudrewicz, Power-Frequency Relations in Nonlinear Two-Poles, *Proceedings of the IRE*, Volume 51, 1963, pp. 1599–1605.
7. C. H. Page, Frequency Conversion with Nonlinear Reactance, *Journal of Research of the NBS*, Volume 58, 1957, pp. 227–236.
8. S. Duinker, General Properties of Frequency-Converting Networks, *Philips Research Reports*, Volume 13, 1958, pp. 37–78, 101–148.
9. H. E. Rowe, Some General Properties of Nonlinear Elements. II. Small Signal Theory, *Proceedings of the IRE*, Volume 46, 1958, pp. 850–860.
10. D. B. Leeson and S. Weinreb, Frequency Multiplication with Nonlinear Capacitors— A Circuit Analysis, *Proceedings of the IRE*, Volume 47, 1959, pp. 2076–2084.
11. K. K. N. Chang, Harmonic Generation with Nonlinear Reactances, *RCA Review*, Volume 19, 1958, pp. 455–464.
12. J. M. Manley, Some General Properties of Magnetic Amplifiers, *Proceedings of the IRE*, Volume 39, 1951, pp. 242–251.
13. E. Della Torre and M. D. Sirkis, Power Conversion with Nonlinear Reactances, *IRE Transactions on Circuit Theory*, Volume CT-8, Number 2, 1961, pp. 95–99.
14. S. Bloom and K. K. N. Chang, Theory of Parametric Amplification Using Nonlinear Reactances, *RCA Review*, Volume 18, 1957, pp. 578–593.
15. B. Salzberg, Masers and Reactance Amplifiers—Basic Power Relations, *Proceedings of the IRE*, Volume 45, 1957, pp. 1544–1545.
16. M. T. Weiss, Quantum Derivation of Energy Relations Analogous to Those for Nonlinear Reactances, *Proceedings of the IRE*, Volume 45, 1957, pp. 1012–1013.
17. J. R. Pierce, Use of the Principles of Conservation of Energy and Momentum in Connection with the Operation of Wave-Type Parametric Amplifiers, *Journal of Applied Physics*, Volume 30, 1959, pp. 1341–1346.

5

LINEARIZATION
AND STABILITY

In several theorems of the last chapter it was asserted that some processes are "potentially unstable" on the basis that the power flowing into a circuit is adequate to overcome losses and the process might sustain itself. This is not a rigorous criterion. In this chapter and subsequent discussions the question of stability will be treated in greater detail. Although the distinction between a network and a system is somewhat moot the material in the remainder of this book is generally associated with nonlinear system analysis and the corresponding notation and nomenclature will be used.

The first part of this chapter presents a rather complete development of an important branch of linear system analysis. The middle portion discusses methods of representing a nonlinear process by various mathematical models. The last section of this chapter presents methods whereby conventional linear system analysis can be extended to include nonlinear networks and systems. Throughout all of this chapter nothing but sinusoidal analysis is used and reference to the third chapter may be helpful.

LINEAR SYSTEM ANALYSIS

Most of the methods of linear system analysis depend to varying degrees on the use of complex variables so the first part of this section is devoted to a review of complex variable theory. Mathematicians traditionally call such a variable z but this discussion leads to the Laplace transform so the variable is $s = \sigma + j\omega$. This is plotted in the s plane with σ along the horizontal axis and $j\omega$ along the vertical axis. The variable may also be represented in the polar form $s = re^{j\theta}$ where r is the radius and θ is measured in a counterclockwise sense from the horizontal axis.

A function $F(s)$ may be formed from s and it may be plotted in its own plane. The derivative of such a function is defined in the usual manner as

$$\frac{dF}{ds} = \lim_{\Delta s \to 0} \frac{\Delta F}{\Delta s} \tag{5.1}$$

but ΔF is a line segment in the F plane so the derivative of a complex function is the ratio of two line segments in different complex planes. It therefore loses the concept of slope which it has in real variable theory. A function is said to be *analytic* at a point if it possesses a derivative at that point and if the derivative has the same value for all possible Δs. To derive a necessary condition one assumes that

$$F(s) = u(\sigma, \omega) + jv(\sigma, \omega) \tag{5.2}$$

and observes that if $\Delta \omega = 0$ and $\Delta \sigma \to 0$ then

$$\lim_{\Delta s \to 0} \frac{\Delta F}{\Delta s} = \frac{\partial u}{\partial \sigma} + j \frac{\partial v}{\partial \sigma} \tag{5.3}$$

If the approach is made parallel to the $j\omega$ axis then $\Delta \sigma = 0$ and $\Delta \omega \to 0$. This gives

$$\lim_{\Delta s \to 0} \frac{\Delta F}{\Delta s} = \frac{\partial u}{j \, \partial \omega} + \frac{\partial v}{\partial \omega} \tag{5.4}$$

For $F(s)$ to be analytic these two limits must be equal. This yields the two requirements

$$\frac{\partial u}{\partial \sigma} = \frac{\partial v}{\partial \omega}$$

$$\frac{\partial v}{\partial \sigma} = -\frac{\partial u}{\partial \omega} \tag{5.5}$$

These are known as the *Cauchy-Riemann equations*. They are also a sufficient condition for the derivative to have the same value for Δs approaching in every direction. This is established by expressing ΔF as a function of $\Delta \sigma$ and $\Delta \omega$. Then the ratio of ΔF to Δs is evaluated as $\Delta s \to 0$. It is found that if the Cauchy-Riemann equations are satisfied then the limit has a unique value.

Before continuing with complex variable theory it is necessary to establish a fundamental result from real variable theory. It is known as *Green's theorem* and it relates a line integral to a surface integral. As shown in Figure 5.1 the curve C is a closed contour of a region R. Between the extremes x_1 and x_2 it is described by $y_1(x)$ on the bottom and $y_2(x)$ on the top. Given a function $P(x, y)$ which has continuous derivatives inside and on C then a

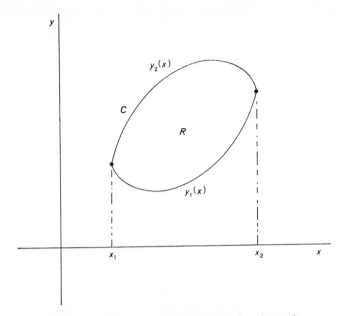

Figure 5.1 Relation between line and surface integrals.

surface integral through R can be expressed as

$$\iint_R \frac{\partial P}{\partial y} \, dx \, dy = \int_{x_1}^{x_2} \left[\int_{y_1}^{y_2} \frac{\partial P}{\partial y} \, dy \right] dx$$

$$= \int_{x_1}^{x_2} [P(x, y_2) - P(x, y_1)] \, dx$$

$$= - \int_C P \, dx \qquad (5.6)$$

For another function $Q(x, y)$ a similar integration between extremes in the y direction gives

$$\iint_R \frac{\partial Q}{\partial x} \, dx \, dy = \int_C Q \, dy \qquad (5.7)$$

and combining both equations gives the final result

$$\int_C [P \, dx + Q \, dy] = \iint_R \left(\frac{\partial Q}{\partial x} - \frac{\partial P}{\partial y} \right) dx \, dy \qquad (5.8)$$

It is now possible to establish the following theorem due to Cauchy which is of great importance in complex variable theory.

Theorem 5.1   *If $F(s)$ is analytic with a continuous derivative inside and on a closed contour then its integral around the contour is zero.*

Applying Green's theorem to a line integral in the complex plane gives

$$\int_C F(s)\, ds = \int_C (u + jv)(d\sigma + j\, d\omega)$$

$$= \int_C [(u\, d\sigma - v\, d\omega) + j(v\, d\sigma + u\, d\omega)]$$

$$= -\iint_R \left(\frac{\partial v}{\partial \sigma} + \frac{\partial u}{\partial \omega}\right) d\sigma\, d\omega$$

$$+ j\iint_R \left(\frac{\partial u}{\partial \sigma} - \frac{\partial v}{\partial \omega}\right) d\sigma\, d\omega \tag{5.9}$$

Since the function is analytic in the region R the Cauchy-Riemann equations are valid and each integrand is zero.

In general, if $F(s)$ is formed from conventional, well behaved expressions the result is analytic and the derivative can be computed from the usual formulas. There are a few important exceptions, however. For example the logarithm is defined as

$$e^{\log s} = s = re^{j\theta}$$
$$\log s = \log r + j\theta \tag{5.10}$$

If the logarithm is computed around a closed path which encloses the origin, when the evaluation returns to the original point it is found that its value has changed by $2\pi j$. An unusual point, such as the origin in this case, about which a function has multiple values is called a *branch point*. (Another example of a branch point is $\sqrt{s}$.) In subsequent discussions it is necessary to consider the logarithm of $F(s)$. It is defined by

$$e^{\log F} = F = Re^{j\phi} \tag{5.11}$$

and the usual rules for differentiation show that

$$d[\log F(s)] = \frac{F'(s)}{F(s)}\, ds \tag{5.12}$$

Branch points are not the only unusual points in the complex plane. Any point at which $F(s)$ fails to be analytic is called a *singular point* or *singularity* of $F(s)$. A singular point is said to be *isolated* if it can be encircled by a curve which encloses no other singular point. If $F(s)$ exists but $F'(s)$ does not then the point is called a *removable singularity*. An example is $s^{-1} \sin s$ at the origin.

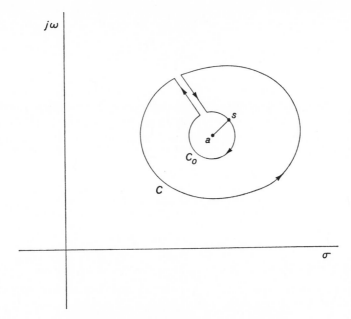

Figure 5.2   Integrating around a simple pole.

If there is a positive integer n such that

$$\lim_{s \to a} (s - a)^n F(s) = c \neq 0 \tag{5.13}$$

then the point $s = a$ is called a *pole*. It has the *order* or *multiplicity n*. These have an important place in complex variable theory due to unusual properties of line integrals which surround them. If an n cannot be found to satisfy Equation 5.13 then there is an *essential singularity*. An example is $e^{1/s}$ at the origin.

To compute the integral around a simple pole where $n = 1$ one considers the integral of an analytic function around the total path shown in Figure 5.2. By Theorem 5.1 this integral must be zero. Hence the integral around the small circle equals the integral around the arbitrary path in the same sense. Since the latter is stationary the integral around the circle must be independent of the radius. If $F(s)$ is analytic then the following integral is also analytic over any portion of the total path and if $s - a = re^{j\theta}$ then

$$\int_{C_0} \frac{F(s)}{s - a}\, ds = \int_{C_0} \frac{F(s)}{re^{j\theta}} jre^{j\theta}\, d\theta$$

$$= j \int_{C_0} F(s)\, d\theta \tag{5.14}$$

As r approaches zero then $F(s)$ approaches the constant value $F(a)$ and this provides an evaluation of the integral around C which is given by

$$F(a) = \frac{1}{2\pi j} \int_C \frac{F(s)}{s-a}\, ds \qquad (5.15)$$

This remarkable result has no analog in real variable theory and is known as *Cauchy's integral formula*. It evaluates a function of a complex variable at a point in terms of a line integral along an arbitrary closed curve surrounding the point. Equally remarkable is the following result which evaluates a derivative at a point.

$$F'(a) = \lim_{\Delta a \to 0} \frac{1}{2\pi j \, \Delta a} \int_C \left[\frac{F(s)}{s-(a+\Delta a)} - \frac{F(s)}{s-a} \right] ds$$

$$= \frac{1}{2\pi j} \int_C \frac{F(s)}{(s-a)^2}\, ds \qquad (5.16)$$

Repeated applications of the above procedure yield the formula

$$\frac{d^n}{ds^n} F(s)\Big|_{s=a} = \frac{n!}{2\pi j} \int_C \frac{F(s)}{(s-a)^{n+1}}\, ds \qquad (5.17)$$

This formula is useful for expanding a function by infinite series. By ordinary long division

$$\frac{1}{s-s_0} = \frac{1}{(s-a)-(s_0-a)}$$

$$= \frac{1}{s-a} + \frac{s_0-a}{(s-a)^2} + \frac{(s_0-a)^2}{(s-a)^3} + \cdots. \qquad (5.18)$$

Multiplying by $F(s)$, integrating along a contour surrounding $s = a$, and replacing s_0 with s give

$$F(s) = F(a) + F'(a)(s-a) + \frac{F''(a)}{2!}(s-a)^2 + \cdots. \qquad (5.19)$$

This is recognized as a Taylor series expansion for $F(s)$ about the point $s = a$. Since the coefficients are derived from Cauchy's integral formula the series does not converge to $F(s)$ if the function is not analytic. If there is a pole at $s = a$ of multiplicity n then $(s-a)^n F(s)$ is analytic and can be expanded by the above procedure to give

$$(s-a)^n F(s) = c_{-n} + \cdots + c_{-1}(s-a)^{n-1}$$

$$+ c_0(s-a)^n + c_1(s-a)^{n+1} + \cdots \qquad (5.20)$$

which can be written in the form

$$F(s) = \frac{c_{-n}}{(s-a)^n} + \cdots + \frac{c_{-1}}{(s-a)} + c_0 + c_1(s-a) + \cdots \quad (5.21)$$

This type of expression is called a *Laurent series*. It is an extension of a Taylor series to express functions which are not analytic. A combination of Equations 5.17, 5.19, and 5.20 shows that the coefficients are given by the formula

$$c_n = \frac{n!}{2\pi j} \int_C \frac{F(s)}{(s-a)^{n+1}} ds \quad (5.22)$$

where C is any closed curve surrounding $s = a$. The coefficient c_{-1} has a very important significance. From the above formula it is observed that

$$\int_C F(s) \, ds = 2\pi j c_{-1} \quad (5.23)$$

so the value of the integral of $F(s)$ around a closed curve enclosing a pole can be found from a single coefficient in the Laurent series. This result is of great practical value. The coefficient c_{-1} is called the *residue* at the pole. In any Laurent expansion containing a pole of highest multiplicity m at $s = a$ the residue can be isolated by performing the operations indicated in the formula

$$c_{-1} = \frac{1}{(m-1)!} \left[\frac{d^{m-1}}{ds^{m-1}} (s-a)F(s) \right]_{s=a} \quad (5.24)$$

and when combined with Equation 5.23 this gives an evaluation of the integral of $F(s)$.

Theorem 5.2 *The integral of a function around a contour enclosing a region equals $2\pi j$ times the sum of the residues of the function in the region.*

This is easily established by assuming that the function is expanded by a Laurent series in the vicinity of each pole. Then the principle illustrated in Figure 5.2 is applied to the case where there are numerous poles within the region.

The next portion of the discussion introduces some new concepts which ultimately will utilize complex variable theory to analyze linear systems. For a given $f(t)$ let there be a real number σ such that

$$\int_{-\infty}^{\infty} |f(t)| \, e^{-\sigma t} \, dt \quad (5.25)$$

exists. Due to the negative values for t it is frequently necessary to define $f(t)$ to be zero to the left of a region known as the *strip of convergence*. The integral

$$F(s) = \int_{-\infty}^{\infty} f(t)e^{-st}\, dt \tag{5.26}$$

certainly converges for the above σ. This defines what is called the *two-sided Laplace transform*. It is a function of s and it can be transformed back to a function related to $f(t)$. This is demonstrated by an examination of the integrals given below.

$$
\begin{aligned}
\frac{1}{2\pi j}\int_{\sigma-j\omega}^{\sigma+j\omega} F(s)e^{st}\, ds &= \frac{1}{2\pi}\lim_{A\to\infty}\int_{-A}^{A} e^{st}\left[\int_{-\infty}^{\infty} f(u)e^{-su}\, du\right] d\omega \\
&= \frac{e^{\sigma t}}{2\pi}\lim_{A\to\infty}\int_{-\infty}^{\infty} f(u)e^{-\sigma u}\left[\int_{-A}^{A} e^{j\omega(t-u)}\, d\omega\right] du \\
&= \frac{e^{\sigma t}}{\pi}\lim_{A\to\infty}\int_{-\infty}^{\infty} f(u)e^{-\sigma u}\frac{\sin A(t-u)}{(t-u)}\, du \\
&= \frac{1}{\pi}\lim_{A\to\infty}\int_{-\infty}^{\infty} f(t+u)e^{\sigma u}\frac{\sin Au}{u}\, du \\
&= \frac{2}{\pi}\lim_{A\to\infty}\int_{0}^{\infty} \bar{f}(t,u)e^{\sigma u}\frac{\sin Au}{u}\, du \tag{5.27}
\end{aligned}
$$

The last integral is in precisely the form of the Dirichlet integral of Equation 3.46 which was used to demonstrate the convergence of Fourier series. In an analogous fashion there is a Riemann localization theorem for the Laplace transform and again it is true that its convergence depends on the behavior of the above integrand around the origin for u. The same method which established Theorem 3.4 can be used to prove the following theorem which is the most fundamental in Laplace transform theory.

Theorem 5.3   *If $f(t)$ is monotonic in sufficiently small regions on both sides of every discontinuity then its two-sided Laplace transform converges everywhere to $f(t)$ except at points of discontinuity where it converges to the average value.*

The direct transform is given by Equation 5.26 and the inverse transform and its convergence are given by

$$\tfrac{1}{2}[f(t^{+}) + f(t^{-})] = \frac{1}{2\pi j}\int_{\sigma-j\omega}^{\sigma+j\omega} F(s)e^{st}\, ds \tag{5.28}$$

where the terms on the left indicate the average value of $f(t)$ calculated by approaching the limit on each side of a discontinuity. Although the two-sided Laplace transform is the more general one it places a severe restriction on $f(t)$. In practical problems it is almost always acceptable to assume that

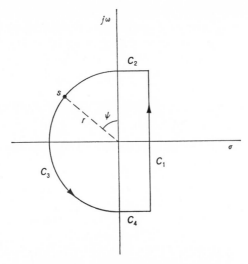

Figure 5.3 A typical Bromwich path.

$f(t) = 0$ for all values of t less than a given t_0. A translation in time places t_0 at the origin. This gives the regular Laplace transform which is described by the equations

$$F(s) = \int_0^\infty f(t)e^{-st}\,dt$$

$$f(t) = \frac{1}{2\pi j}\int_{\sigma-j\omega}^{\sigma+j\omega} F(s)e^{st}\,ds$$

(5.29)

This transform has the same convergence properties provided that $|f(t)|$ converges over the infinite interval with an exponential weighting factor. The following are the two most useful Laplace transform pairs.

$$f(t) = t^n \leftrightarrow F(s) = \frac{n!}{s^{n+1}}$$

$$f(t) = e^{j\omega t} \leftrightarrow F(s) = \frac{1}{s - j\omega}$$

(5.30)

The transforms of the step function and the ramp function frequently used in linear system analysis are obtained from the first expression by letting $n = 0$ and $n = 1$ respectively. The transforms of sine and cosine waves are found from a linear combination of two equations derived from the second expression by considering positive and negative exponents.

The inverse transform is conveniently calculated using the theory of residues. The usual contour is shown in Figure 5.3. It is known as a *Bromwich path*. Although σ may never acquire a value on the vertical line which is below the minimum required for convergence of the direct transform it may be as large as desired. It is therefore made sufficiently great to assure that all singularities lie to the left of the vertical line. To apply the theory of residues it is necessary to assume that $F(s)$ has the property that

$$F(s) < M(r) \tag{5.31}$$

where M is an upper bound applicable for all values of ψ on the curves and where $M \to 0$ as $r \to \infty$. Using the general principle that

$$\int_{s_1}^{s_2} F(s)e^{st}\,ds \leqslant \int_{s_1}^{s_2} |F(s)|\,|e^{st}|\,|ds| \tag{5.32}$$

it is immediately apparent that the integrals along C_2 and C_4 approach zero as $r \to \infty$ because they involve a real differential and a vanishing integrand. Since $|ds| = r\,d\psi$ along curve C_3 the following relations are true

$$\int_{C_3} F(s)e^{st}\,ds \leqslant \int_0^\pi Me^{-tr\sin\psi}r\,d\psi$$

$$< 2rM \int_0^{\pi/2} e^{-(2tr\psi)/\pi}\,d\psi$$

$$= \frac{M\pi}{t}(1 - e^{rt}) \leqslant \frac{M\pi}{t} \tag{5.33}$$

As $r \to \infty$ the integral around C_3 also approaches zero. Therefore the sum of the residues within the Bromwich path evaluates the line integral along the remaining curve C_1. This is fortunately the path required for the inverse transform. Following is an example to illustrate the use of this principle.

A typical transform arising from the analysis of a linear system may have the form

$$F(s) = \frac{8}{s(s+2)}\frac{1}{s} \tag{5.34}$$

This has a simple pole at $s = -2$ and a double pole at $s = 0$. The residue at the simple pole is found by the formula to be

$$c_{-1}\big|_{s=-2} = (s+2)F(s)e^{st}\,\big|_{s=-2}$$

$$= \frac{8e^{st}}{s^2}\bigg|_{s=-2} = 2e^{-2t} \tag{5.35}$$

and the residue at the double pole is given by

$$c_{-1}|_{s=0} = \frac{d}{ds} s^2 F(s) e^{st} \Big|_{s=0}$$

$$= \frac{d}{ds} \frac{8 e^{st}}{s+2} \Big|_{s=0}$$

$$= 8 \left[\frac{t}{s+2} - \frac{1}{(s+2)^2} \right] e^{st} \Big|_{s=0}$$

$$= 4t - 2 \tag{5.36}$$

From the above the Laurent expansion for $F(s)$ and the solution for $f(t)$ are

$$F(s) = \frac{2}{s+2} + \frac{4}{s^2} - \frac{2}{s}$$

$$\tag{5.37}$$

$$f(t) = 2e^{-2t} + 4t - 2$$

Up to this point the Laplace transform appears as a method for converting a function of time into a function of a complex variable and then converting it back again. Since the interval in the time domain is infinite and the frequency spectrum in the complex domain is continuous it represents a considerable extension of a Fourier series analysis but its application to linear system analysis has yet to be explained. This is based on another rather unusual and perhaps unexpected property of the transform.

If the direct transform given in Equation 5.29 is evaluated by an integration by parts

$$\int_0^\infty f(t) e^{-st} dt = -\frac{1}{s} f(t) e^{-st} \Big|_0^\infty + \frac{1}{s} \int_0^\infty f'(t) e^{-st} dt$$

$$\tag{5.38}$$

$$\int_0^\infty f'(t) e^{-st} dt = sF(s) - f(0)$$

then a new property emerges. If one forms the time derivative of $f(t)$ and assumes zero initial conditions then its transform is $sF(s)$. Forming a derivative in the time domain is equivalent to a multiplication by s in the complex variable domain. A similar integration by parts shows that an integration in the time domain is equivalent to a division by s. These properties permit the Laplace transform to be applied to the solution of integro-differential equations.

In the analysis of linear systems one frequently meets differential equations of the very general form

$$A_n y^{(n)} + \cdots + A_2 \ddot{y} + A_1 \dot{y} = f(t) \tag{5.39}$$

Taking the Laplace transform of the equation term by term and ignoring initial conditions give

$$[A_n s^n + \cdots + A_2 s^2 + A_1 s] Y(s) = F(s) \qquad (5.40)$$

which can be written in the form

$$Y(s) = \frac{1}{A_n s^n + \cdots + A_2 s^2 + A_1 s} F(s) \qquad (5.41)$$

This fundamental equation states that the transform of the required $y(t)$ is the product of a function of s which incorporates the properties of the system and a function of s which is the direct transform of the forcing function, excitation, or input $f(t)$. The first function of s is frequently called the *transfer function*. The solution for the response or output $y(t)$ is found by evaluating the inverse transform of the product of the two functions.

In networks of electrical devices and more complex physical systems the transfer function is found to be a fraction, the ratio of two polynomials in s. In the closed-loop system shown in Figure 5.4 each of the transfer functions G and H should be considered as fractions containing polynomials in s. From the fundamental relation

$$G(u - Hv) = v \qquad (5.42)$$

there is immediately derived the *closed-loop transfer function*

$$\frac{v}{u} = \frac{G}{1 + GH} \qquad (5.43)$$

The product GH contains all of the information about the system but it does not include the effect of the loop closure. It is called the *open-loop transfer function*.

It is a consequence of the manner in which connections must be made between sources and sinks (for example, series inductance and shunt capacitance in electrical networks) that the numerator of the open-loop transfer function can never be of higher order than the denominator. The function has

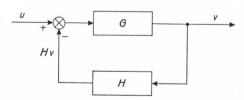

Figure 5.4 Block diagram of the elementary closed-loop system.

the general form

$$GH = K' \frac{s^m + \cdots + a_0'}{s^n + \cdots + b_0} \tag{5.44}$$

where $m \leqslant n$. Upon clearing fractions it is found that the closed-loop transfer function has the general form

$$1 + GH = K \frac{s^n + \cdots + a_0}{s^n + \cdots + b_0} \tag{5.45}$$

where $K = 1$ if $m < n$. This establishes the interesting fact that the denominator of the closed-loop transfer function has equal numbers of poles and zeros.

In systems possessing distributed parameters such as transmission lines or particles passing through variable fields the transfer function may contain transcendental terms such as hyperbolic functions. In these cases the transfer function can be expanded as a Laurent series and the inverse transform again depends on the nature of the residues.

The above brief outline shows how the Laplace transform can be used to determine the response of a system to a general input. The output appears as a combination of the characteristics of the system and the nature of the input signal. When considering only the stability of the system the exact form of the input signal is irrelevant. Since the system is linear the principle of superposition applies and the stability is independent of the input. If, however, one requires $u(t) = 0$ in the closed-loop system then the system remains quiescent forever and nothing is learned about its own personality. This problem is solved very simply in the s domain by requiring $U(s) = 1$. Then the inverse transform for $v(t)$ reflects exactly the properties of the closed-loop transfer function. But what kind of a function of time has a Laplace transform equal to unity? If one considers it as the product of s and $1/s$ then in the time domain it is the derivative of the step function. This peculiar concept is called the *delta function*. It appears as a sharp spike of unit area appearing at $t = 0$. It gives the system a unit impulse and then allows it to display its own characteristics independently for all $t > 0$.

A system will be considered stable if it tends to return to its original condition after receiving the unit impulse. Due to the factor e^{st} in the integrand for the inverse transform the system response to the impulse contains a sum of terms each containing an exponential factor for all poles except a pole at the origin. Therefore it is necessary and sufficient for stability that (a) all poles of the closed-loop transfer function must lie in the left half of the s plane and (b) a pole at the origin must have a finite residue.

Fortunately mathematicians have found several methods for solving this problem. Poles for the closed loop transfer function are zeros for $1 + GH$.

If G and H are in polynomial form then the problem becomes that of determining whether or not the roots of a polynomial have negative real parts. Classical methods for doing this are due to Hurwitz and Routh. A newer and simpler method due to Wall and derived from a study of continued fractions is given here without proof. For the equation

$$s^3 + 7s^2 + 14s + 8 = 0$$

the method involves the following algorithm:

$$
\begin{array}{cccccccc}
 & & & \tfrac{1}{7} & 1 & & & \\
7 & 0 & 8 & \big| 1 & 7 & 14 & 8 & \\
 & & & 1 & 0 & \tfrac{8}{7} & & \\
\hline
 & & & 7 & \tfrac{90}{7} & 8 & & \\
 & & & 7 & 0 & 8 & \tfrac{49}{90} & \\
\hline
 & & & & \tfrac{90}{7} & 0 & \big| 7 & 0 & 8 \\
 & & & & & 7 & 0 & & \tfrac{90}{56} \\
\hline
 & & & & & 0 & 8 & \big| \tfrac{90}{7} & 0 \\
 & & & & & & & \tfrac{90}{7} & \\
\hline
 & & & & & & & 0 & 0 \\
\end{array}
$$

The original coefficients are the first dividend. The first divisor is formed by starting with the next to leading term and replacing alternate coefficients with zero. Each successive remainder becomes a new divisor into the previous divisor. The polynomial has roots with negative real parts if and only if all quotients are positive. The above example represents a stable system.

Only when the system is described by a linear differential equation with constant coefficients does the denominator of its transfer function have such a simple form. To treat more general cases and also to provide a method which can be extended to nonlinear systems it is necessary to present a more powerful technique. To do this, however, it is first necessary to introduce another theorem from complex variable theory.

Theorem 5.4 *As s traverses a closed contour in the s plane in the positive sense then the number of positive encirclements of its origin by $F(s)$ gives the excess of the total multiplicity of zeros over poles of $F(s)$ in the region bounded by the contour.*

This is established in the general case where $F(s)$ is not a factorable polynomial but at least is analytic inside and on the contour except for a zero of order z at $s = a$ and a pole of order p at $s = b$. Let C_1 be a small circle around a. Then around C_1 the function $F(s)$ can be written as

$$F(s) = (s - a)^z R(s) \qquad (5.46)$$

where $R(s)$ is analytic and never zero inside or on C_1. Taking the logarithm of both sides of the equation and differentiating give

$$\frac{F'(s)}{F(s)} = \frac{z}{s-a} + \frac{R'(s)}{R(s)}$$ (5.47)

and integrating around C_1 where $s - a = re^{j\theta}$ gives

$$\frac{1}{2\pi j} \int_{C_1} \frac{F'(s)}{F(s)} \, ds = \frac{z}{2\pi j} \int_{C_1} \frac{ds}{s-a} + \frac{1}{2\pi j} \int_{C_1} \frac{R'(s)}{R(s)} \, ds$$

$$= \frac{z}{2\pi j} \int_{C_1} j \, d\theta$$

$$= z$$ (5.48)

Similarly, if C_2 is a small circle around b then $F(s)$ can be written as

$$F(s) = \frac{S(s)}{(s-b)^p}$$ (5.49)

where $S(s)$ is analytic and never zero inside and on C_2. Taking the logarithm and differentiating now give

$$\frac{F'(s)}{F(s)} = \frac{S'(s)}{S(s)} - \frac{p}{s-b}$$ (5.50)

Integrating around C_2 gives

$$\frac{1}{2\pi j} \int_{C_2} \frac{F'(s)}{F(s)} \, ds = -p$$ (5.51)

Applying the principle shown in Figure 5.2 to this case where there are two small circles shows that for a contour C surrounding both circles

$$\frac{1}{2\pi j} \int_C \frac{F'(s)}{F(s)} \, ds = z - p$$ (5.52)

At the same time if $F(s) = Re^{j\phi}$ in the F plane then applying Equation 5.12 and integrating around a closed contour give

$$\frac{1}{2\pi j} \int_C \frac{F'(s)}{F(s)} \, ds = \frac{1}{2\pi j} \int_C d[\log F(s)]$$

$$= \frac{1}{2\pi j} \int_C [d(\log R) + j \, d\phi]$$

$$= N$$ (5.53)

where N is the number of positive encirclements of the origin by $F(s)$ as s traverses the contour C in the s plane. Combining both evaluations of the integral and using the principle shown in Figure 5.2 to extend the contour to enclose many poles and zeros give the final result

$$N = Z - P \qquad (5.54)$$

where Z and P are the total multiplicities of all zeros and poles enclosed by C. For example, if C encloses one double zero and two simple zeros then $Z = 4$.

Since the determination of the stability of a linear system depends on a verification of the existence of zeros in the right half plane the applicability of Theorem 5.4 is immediately apparent. For the various paths shown in part (a) of Figure 5.5 the corresponding plots of $1 + GH$ may have the appearance shown in part (b). As the radius of the circle increases, the contour $OACBO$ eventually encloses all poles and zeros in the right half plane and the behavior of the plot of $1 + GH$ betrays their existence. When a consistent sense is assigned to the paths in Figure 5.5(b) it is concluded that $1 + GH$ has two excess zeros within the contour if all poles and zeros are simple. Thus the system is unstable.

The path $OAC'BO$ encircles the origin an equal number of times. Due to its sense it shows that $1 + GH$ has two excess poles in the left semicircle. This symmetry is due to the previously stated fact that $1 + GH$ has equal numbers of poles and zeros in the entire s plane.

If $1 + GH$ has the form shown in Equation 5.45 then as the size of the circle is increased the leading terms predominate and $1 + GH$ approaches the real number K. In any realizable physical system the transmission ultimately falls to zero if the frequency is increased sufficiently. This means that any realistic GH has more poles than zeros and in Equation 5.44 $m < n$ and therefore $K = 1$. When the circle is increased in the s plane its counterpart in the $1 + GH$ plane coalesces to the point $s = 1$. Thus the circular paths can be ignored and all information is contained in the plot along the imaginary axis.

A further simplification is possible. Regardless of the manner in which they are expressed the only difference between GH and $1 + GH$ is a lateral translation of one unit. It is simpler to plot GH and to consider encirclements of the point $s = -1$ rather than the origin. The point $(-1, 0)$ is frequently called the *critical point*.

This procedure was first presented by Nyquist in 1932. It was derived to show the feasibility of the use of feedback in amplifiers used in the telephone system. As a matter of historical interest the following statement is taken almost verbatim from his original report.

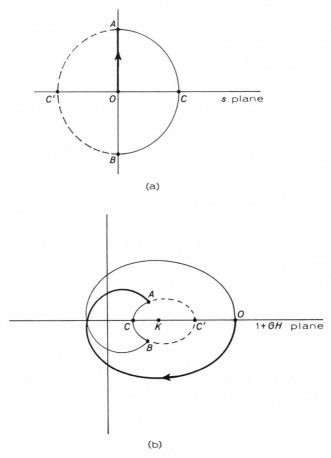

(a)

(b)

Figure 5.5 (a) The s plane contours. (b) The paths for $1 + GH$.

Rule: Plot plus and minus the imaginary part of GH against the real part for all frequencies from 0 to ∞. If the point −1 + j0 lies completely outside this curve the system is stable; if not it is unstable.

Further simplification is possible. Since the original differential equation for the system contains only real coefficients the plot of GH for negative frequencies is the conjugate of the plot for positive frequencies. Hence it is customary to plot $G(j\omega)H(j\omega)$ for values of ω only from 0 to $+\infty$. A factor of two must be introduced into the interpretation of Theorem 5.4.

It should be remembered that the behavior of the plot indicates only the *excess* of zeros in the right half plane. If G contains a closed loop then GH may have poles in the right half plane. For such multiple loop systems it is convenient to apply the Nyquist criterion successively, starting with the innermost loop.

The calculation of a Nyquist plot consists of substituting $j\omega$ for s in the transfer function. Taking a time derivative of a sinusoid advances the phase 90° and multiplies by ω. This is equivalent to multiplying by $j\omega$ in complex notation. As a result the polar plot of GH is the amplitude and phase response of the system as a complex function of ω. This can be determined experimentally. This combination of mathematical simplicity and direct relationship to the physical system is an important feature of the Nyquist technique.

Prior to the development of the Nyquist theory it was thought that if sufficient feedback is present at a phase shift of 180° then the system would overcome losses in its input circuit and would necessarily be unstable. The Nyquist theorem, however, shows that it is possible for the plot of GH to intersect the real axis to the left of the critical point at $(-1, 0)$ and yet the system can be stable. This puzzling effect is called *conditional stability* since a decrease in open-loop gain can result in encirclement of the critical point. A typical plot is shown in Figure 5.6. This shows that stability depends on the characteristic of the open-loop transfer function for many frequencies rather than just those producing 180° phase shift.

Theorem 5.5   *If the denominator of the open-loop transfer function is no higher than second order and if the numerator is no higher than first order then the system is stable.*

It is assumed that the feedback is in the proper sense so for zero frequency the Nyquist plot terminates on the positive real axis. (If there is a pole at the

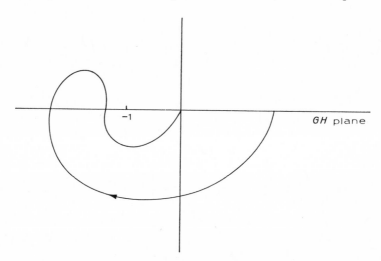

Figure 5.6   Nyquist plot showing conditional stability.

origin then it must be avoided by the method shown in the following example.) If both numerator and denominator are of first order then the maximum phase shift at any frequency is 90° so the plot of GH never reaches the left half plane. If the denominator is of second order then a phase lag of 180° may occur at infinite frequency. However, since the numerator is of lower order the amplitude approaches zero at infinite frequency so the plot of GH terminates at the origin and therefore cannot contact the critical point. The following examples demonstrate this theorem and other aspects of the Nyquist method.

The first example assumes direct feedback so $H = 1$. The transfer function in the forward portion is given by

$$G(s) = \frac{1}{s(2s + 1)} \tag{5.55}$$

Along the imaginary axis this becomes

$$G(j\omega) = \frac{-2\omega^2 - j\omega}{4\omega^4 + \omega^2} \tag{5.56}$$

This function has a pole at the origin. This represents the capability of pure integration and is typical of devices such as electric motors. Application of the Nyquist theorem does not permit the s plane contour to pass through poles or zeros. This is avoided by taking a detour as shown in Figure 5.7(a). The pole is arbitrarily placed to the left of the s plane contour and is ignored in the determination of the excess zeros in the right half plane. To perform the Nyquist plot a radius of 0.25 is arbitrarily selected so along the circular portion s and $G(s)$ have the forms

$$s = 0.25e^{j\theta} \tag{5.57}$$

$$G(s) = \frac{1}{0.125e^{j2\theta} + 0.25e^{j\theta}}$$

As s follows the specified contour the use of the functions specified in Equations 5.56 and 5.57 produces the Nyquist plot shown in Figure 5.7(b). This curve does not enclose the $(-1, 0)$ point and the open-loop transfer function has no right half plane poles. It is concluded therefore that the closed-loop system has no right half plane zeros and is stable. Since $G(s)$ is only quadratic it is possible to verify this by calculating the roots of $1 + G$ directly. These are found to lie at $s = -1 \pm j\sqrt{7}$ and are in the left half plane. The two poles of $1 + G$ also lie to the left of the s plane contour at $s = 0$ and $s = -0.5$.

The second example also uses direct feedback and has the open-loop transfer function

$$G(s) = \frac{s - 3}{(s + 1)(s + 2)} \tag{5.58}$$

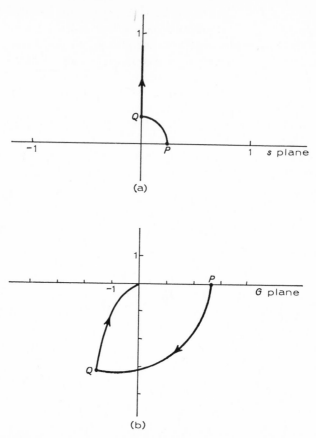

Figure 5.7   (a) The s plane contour. (b) The Nyquist plot.

There are no poles or zeros on the imaginary axis so the standard s plane contour may be employed. However, to demonstrate the complete theory the contour shown in Figure 5.8(a) is used. The resultant Nyquist plot is shown in Figure 5.8(b). For simplicity the conjugate path is not drawn but it can easily be visualized as an image reflected across the real axis. In particular the point B is the image of point A. With this visualization the curve $OACBO$ encircles the origin and the point $(-1, 0)$ each one time in a clockwise sense. This shows the existence of one excess right half plane zero in both G and $1 + G$. The curve $OAC'BO$ traverses the s plane in the positive sense so clockwise encirclements indicate poles rather than zeros. The plot encircles the origin twice and the critical point once. This indicates two excess left half plane poles for G and one for $1 + G$. A direct solution shows that $1 + G$ has zeros at $s = 0.24$ and $s = -4.24$. Its poles of course lie at $s = -1$ and

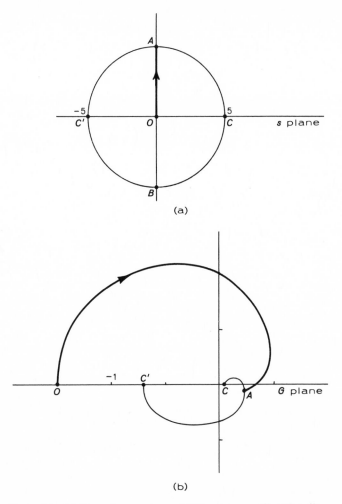

Figure 5.8   (a) The *s* plane contours. (b) The corresponding Nyquist plots.

$s = -2$. Simple enumeration verifies the graphical determinations. This system is unstable due to the right half plane zero indicated by the passage of the path from O to A one half of the way around the critical point.

If a system employs direct feedback making $H = 1$ then its closed-loop transfer function is reduced to

$$\frac{v}{u} = \frac{G}{1 + G} \tag{5.59}$$

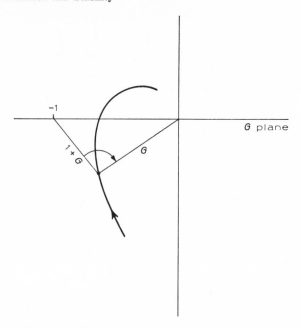

Figure 5.9 Closed-loop response from the Nyquist plot.

This may be evaluated easily from a Nyquist plot as shown in Figure 5.9. For a given frequency it is necessary only to divide the lengths of the vectors to get the amplitude response and to subtract angles to obtain the phase response. A simple geometric consideration shows that the indicated angle between the vectors is actually the difference between the angles of the vectors and the phase response may be read directly. Passage close to the critical point reduces the length of the $1 + G$ vector and produces a peak in the amplitude response. If there is a frequency sensitive element in the feedback path then the Nyquist plot involves the total transfer characteristic GH. The above method may still be used but it is necessary to divide the results by $H(j\omega)$ to obtain the output response.

LINEARIZATION

Since there are convenient engineering techniques available for the analysis and design of linear systems it is always tempting to use these methods as an approximate means of analyzing nonlinear systems. In order to use them the nonlinear process must be represented in a linear fashion. The search for the optimum form of such representation has been the subject of considerable study and it appears that there is no universally valid criterion. Linearization is a form of approximation which must be used with discretion and the

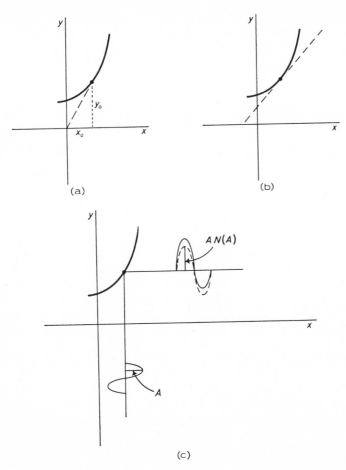

Figure 5.10   (a) Static linearization. (b) Dynamic linearization. (c) The describing function.

results are almost always subject to doubt. What is most disconcerting is the fact that it is usually quite difficult to perform an auxiliary calculation to verify the accuracy of the approximation. Regardless of these difficulties the designer faced with the problem of analyzing a nonlinear system has no choice other than to use the methods which are available and linearization will provide a result in all cases.

Three methods are shown in Figure 5.10. In what might be called "static linearization" the transfer characteristic is represented as the ratio of y_0 to x_0 which occurs at the assumed operating point. This is probably the most crude approximation which is possible and is used only in rough calculations.

A slightly more refined representation is shown in part (b) of Figure 5.10. This might by contrast be called "dynamic linearization". It uses the slope of the characteristic curve at the operating point. This method can be refined further by averaging the slope over the anticipated interval.

An excellent analysis of various methods of linearization has been made by Krylov and Bogoliubov. A summary of their work by Lefschetz is listed as a reference for this chapter. Their analysis is based on the differential equation of a second-order system with a force function containing a weak nonlinear term. A solution is assumed wherein the amplitude and frequency of the oscillation are slowly varying functions of time. Various techniques are introduced to solve the equation within specified limits of accuracy. In one method the average values of certain components of the oscillatory force function are used to determine the time dependency of the rate of change of the amplitude and frequency. This method is called the "first approximation".

A solution which has higher accuracy is obtained by defining terms in a tractable linear equation such that the fundamental components of the oscillation of the force function in the nonlinear system and the corresponding linear system are the same. This is called the principle of "harmonic balance".

This and other intuitive considerations have led to the representation known as the "describing function". The development of the describing function occurred somewhat simultaneously with different workers in several countries and consequently it is impossible to ascribe unique authorship. The basic principle is illustrated in Figure 5.10(c). A sinusoidal input with amplitude A is assumed and the describing function $N(A)$ is defined as the *complex ratio* of the amplitude of the *fundamental component* of the output to the amplitude of the input. Due to the nonlinearity this ratio depends on A. If the nonlinearity is single valued then the methods of interpolation and harmonic generation discussed in the third chapter show that there is no phase shift and the describing function is a real function. If there is hysteresis then it will be complex. If there are characteristics such as a constant time delay or velocity limiting then the describing function also depends on the frequency.

The general procedure for calculating the describing function is to assume an input wave such as

$$x = A \cos \omega t \tag{5.60}$$

and then to compute the amplitude of the fundamental component in $y(x)$ by a Fourier type of analysis. The result is divided by A to get the gain at the fundamental frequency. This process is expressed as

$$N(A) = \frac{1}{\pi A} \int_{-\pi}^{\pi} y(A \cos \theta) e^{-j\theta} \, d\theta \tag{5.61}$$

A large part of Chapter 3 is devoted to sinusoidal analysis and several methods including graphical techniques are described which provide an evaluation of the above integral. It should perhaps be mentioned that Douce has developed a graphical method which facilitates the direct solution using the Lewis technique. His report is listed as a reference and his method is recommended for situations requiring repeated graphical solutions.

Figure 5.11 shows the describing function for a few basic nonlinearities. The double-valued function is a crude simulation of hysteresis and shows the typical phase shift. No attempt is made to provide a complete tabulation of common describing functions since these have been presented many times in the literature. And should such reference material not be available or should the particular nonlinearity not be listed then the methods presented in the third chapter can be used. If the nonlinearity is described by experimental information then graphical methods or numerical integration may be necessary. If the nonlinear characteristic can be expressed as the sum of several functions whose describing functions are known then the describing function of the sum equals the sum of the describing functions.

There is a property of a Fourier series which gives additional insight into the accuracy of Fourier approximations and which may assist in applying the describing function. Although it is not immediately apparent, the formulas for the Fourier coefficients provide a trigonometric interpolation which minimizes the mean square error. Considering an arbitrary sum of the form

$$s_n = \sum_{k=-n}^{n} A_k e^{jkt} \qquad (5.62)$$

then the mean square error between this sum and the given function $f(t)$ is expressed by

$$\Delta_n = \frac{1}{2\pi} \int_{-\pi}^{\pi} [f(t) - s_n(t)]^2 \, dt \qquad (5.63)$$

For each complex amplitude to be selected so as to minimize the mean square error it is necessary that

$$\frac{\partial \Delta_n}{\partial A_k} = 0 = -\frac{1}{\pi} \int_{-\pi}^{\pi} [f(t) - s_n(t)] \frac{\partial s_n}{\partial A_k} \, dt$$

$$= -\frac{1}{\pi} \int_{-\pi}^{\pi} [f(t) - s_n(t)] e^{jkt} \, dt \qquad (5.64)$$

The orthogonality between $s_n(t)$ and the exponential function reduces this expression to

$$A_{-k} = \frac{1}{2\pi} \int_{-\pi}^{\pi} f(t) e^{jkt} \, dt \qquad (5.65)$$

This is the formula by which Fourier coefficients are calculated. Since this result is independent of n it shows that a Fourier sum minimizes the mean square error for any number of terms. This is true for $n = 1$ and therefore is a property of the describing function.

Perhaps the most penetrating effort to establish a firm mathematical foundation for the use or abandonment of the describing function has been made by Bass. He has established several necessary and sufficient conditions

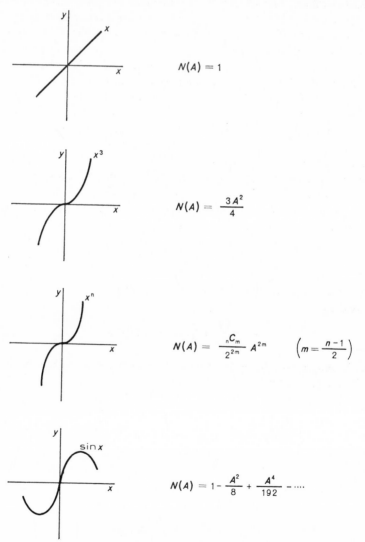

$$N(A) = 1$$

$$N(A) = \frac{3A^2}{4}$$

$$N(A) = \frac{{}_nC_m}{2^{2m}} A^{2m} \qquad \left(m = \frac{n-1}{2} \right)$$

$$N(A) = 1 - \frac{A^2}{8} + \frac{A^4}{192} - \cdots$$

Figure 5.11 Some basic describing functions.

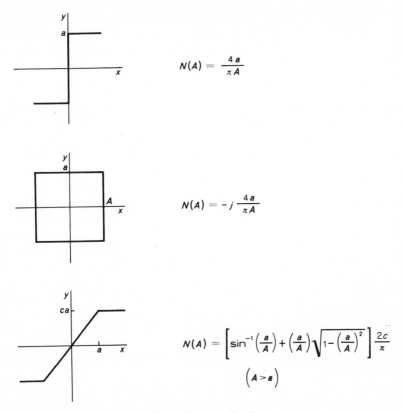

$$N(A) = \frac{4a}{\pi A}$$

$$N(A) = -j\,\frac{4a}{\pi A}$$

$$N(A) = \left[\sin^{-1}\left(\frac{a}{A}\right)+\left(\frac{a}{A}\right)\sqrt{1-\left(\frac{a}{A}\right)^2}\,\right]\frac{2c}{\pi}$$

$$\left(A > a\right)$$

Figure 5.11 continued.

for the existence in a given dynamical system of periodic motion described as "π symmetric." As the name suggests these are undamped vibrations which possess odd symmetry relative to an angular translation of π radians. The use of the describing function tacitly assumes vibrations of only this type. Although the results of Bass are quite complete and mathematically satisfying they are difficult to apply. The criteria involve either the solution of simultaneous equations incorporating Fourier type integrals of the non-linearities or the determination of certain relationships between the actual system and a hypothetical, related system whose properties are known. Either of these criteria impose great computational difficulties when there are several degrees of freedom. Also, the analysis is restricted to second order systems wherein the force function is odd. Although only odd functions within the system can generate signals at the fundamental frequency the presence of even functions can change the operating mode. A comprehensive report by Bass is listed as a reference, not as a source for solutions of practical problems but as an indication of the extreme difficulty in establishing the legitimacy of the describing function even when very elegant and abstract mathematical techniques are employed. However, it is concluded that the describing function is a reliable tool for analyzing most commonly occurring systems and a considerable amount of experimental evidence confirms this opinion.

LINEARIZED SYSTEM ANALYSIS

In the preceding section the concept of the describing function is developed and it is presented as probably the most refined general method for linearization. However, the describing function is a function primarily of amplitude and the transfer functions of the linear portion of the system are functions of frequency. It might appear that these two methods of expression are in conflict and this would prohibit any standardization of procedure. Fortunately this is not true. It is possible to develop a simple block diagram which is representative of a large class of nonlinear systems and to derive a simple, direct procedure for predicting the properties of the system using an interesting modification of the Nyquist technique. First, however, it is necessary to investigate permissible changes in configuration.

Stout has summarized the possible changes in a block diagram which includes one nonlinear element. The procedures become very restricted when there are several separated nonlinear elements. Although his report is listed as a reference the most pertinent result will be repeated here. It is assumed that a nonlinear element is contained in an otherwise linear network in the manner shown in Figure 5.12(a). Using the principle of superposition in the linear portion permits the representation by the general block diagram shown in Figure 5.12(b) where the elements are defined by the relations

$$y_{21} = \left.\frac{i_2}{v_1}\right|_{v_2=0}$$

$$y_{22} = \left.\frac{i_2}{v_2}\right|_{v_1=0}$$

$$A_{31} = \left.\frac{v_3}{v_1}\right|_{v_2=0}$$

$$A_{32} = \left.\frac{v_3}{v_2}\right|_{v_1=0}$$

(5.66)

It should be observed that the block diagram not only provides the proper output voltage but also the proper excitation to the nonlinear element when the input voltages are the same.

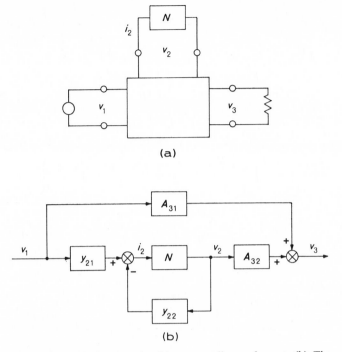

Figure 5.12 (a) A network with one nonlinear element. (b) The equivalent block diagram.

From these considerations it appears that an extremely general situation can be represented by the block diagram in Figure 5.13(a). By observing that

$$x = G_3(e - G_4 r) - G_2 v$$
$$r = G_5 v$$

(5.67)

it is recognized that an equivalent block diagram is shown in Figure 5.13(b). Except for linear operations on the excitation and the response this may be further reduced to the form shown in Figure 5.13(c). This is a very important result and a great part of the analytical work that has been done on linearized systems is based on this block diagram. The closed loop transfer function is

$$\frac{N(A)G(s)}{1 + N(A)G(s)}$$

(5.68)

and stability depends on the nature of the roots of the equation

$$1 + N(A)G(s) = 0$$

(5.69)

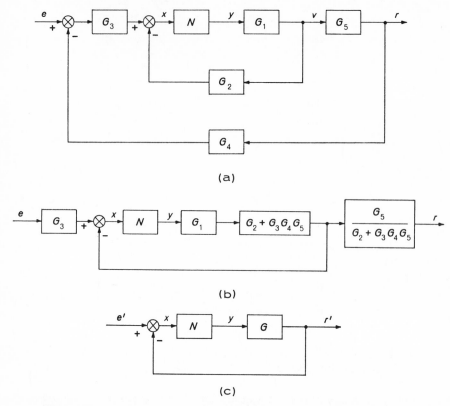

Figure 5.13 Simplification of a very general system with one nonlinear element.

The use of the Nyquist technique is immediately suggested. However, for each amplitude to the nonlinear element $N(A)$ has a different value and the single Nyquist plot for a linear system becomes a family of curves for the linearized system. This difficulty is cleverly avoided by writing Equation 5.69 in the form

$$\frac{1}{N} + G = 0 \qquad (5.70)$$

This seemingly trivial change offers an entirely new approach. The Nyquist critical point at $(-1, 0)$ is due only to the lateral translation by one unit to convert a plot of G into a plot of $1 + G$. Examination of Equation 5.70 shows that the translation is now a variable represented by $1/N$. The critical point now becomes a critical locus and the family of NG plots can be replaced by only two curves. One curve shows G as a function of frequency and the other shows $-1/N$ as a function of amplitude. For any given input amplitude the standard Nyquist criterion may be applied. Furthermore, if both curves pass through the same point then this gives values for A and ω which satisfy Equation 5.69. This indicates an undamped oscillation at that amplitude and frequency. Although it is impossible to maintain an undamped oscillation in a linear system it is quite feasible in a nonlinear system due to the existence of $N(A)$. Such a mode of performance is called a *limit cycle*. Frequently nonlinear networks or systems are employed specifically for the generation of steady oscillations. In this application the definition of stability must be completely revised. Satisfactory performance requires that A and ω remain nearly constant. A stable limit cycle is defined as one which returns to its original value after a temporary external disturbance. An unstable limit cycle is one wherein a disturbance causes the oscillation either to decrease to zero or change to another mode.

A systematic study of the requirements for a stable limit cycle has been made by Loeb and one of his more comprehensive reports is listed as a reference. His analysis considers the very general case wherein it is impossible to separate the amplitude sensitive and frequency sensitive portions of the open-loop transfer function. This is the case if the describing function is frequency sensitive. For such a situation the basic equation may be written as

$$1 + H(A, \omega) = 0 \tag{5.71}$$

If there is an external disturbance then a solution of the form

$$x = Ae^{j\omega t} \tag{5.72}$$

would in general experience a slight change in amplitude and frequency. If the solution has any property of equilibrium then it must also acquire damping as it departs from the equilibrium condition. The new solution would have the form

$$x = (A + \delta A)e^{j(\omega + \delta\omega - j\epsilon)t} \tag{5.73}$$

where ϵ indicates the damping. The basic equation now becomes

$$1 + H(A + \delta A, \omega + \delta\omega - j\epsilon) = 0 \tag{5.74}$$

Expanding this by the first term in a Taylor series gives

$$\frac{\partial H}{\partial A}\delta A + \frac{\partial H}{\partial \omega}(\delta\omega - j\epsilon) = 0 \tag{5.75}$$

For this expression to be valid it is necessary that H is analytic with respect to A and ω. It is also implied that the change in frequency is so slow that sideband power remains in the passband of the system and effects due to changing frequency may be ignored.

Since $H(A, \omega)$ is in general complex it can be represented as

$$H = U + jV \tag{5.76}$$

and the real and imaginary terms provide the two equations

$$\frac{\partial U}{\partial A} \delta A + \frac{\partial U}{\partial \omega} \delta \omega + \frac{\partial V}{\partial \omega} \epsilon = 0$$
$$\frac{\partial V}{\partial A} \delta A + \frac{\partial V}{\partial \omega} \delta \omega - \frac{\partial U}{\partial \omega} \epsilon = 0 \tag{5.77}$$

Eliminating $\delta \omega$ between these gives

$$\left[\left(\frac{\partial U}{\partial \omega} \right)^2 + \left(\frac{\partial V}{\partial \omega} \right)^2 \right] \epsilon = - \left[\frac{\partial U}{\partial A} \frac{\partial V}{\partial \omega} - \frac{\partial V}{\partial A} \frac{\partial U}{\partial \omega} \right] \delta A \tag{5.78}$$

For a stable limit cycle an increase in amplitude should introduce negative damping. The converse is also true, so for stability it is necessary and sufficient that ϵ and δA have opposite sign. This happens if

$$\frac{\partial U}{\partial A} \frac{\partial V}{\partial \omega} - \frac{\partial V}{\partial A} \frac{\partial U}{\partial \omega} > 0 \tag{5.79}$$

This is equivalent to the requirement

$$\text{Im} \frac{\partial H^*}{\partial A} \frac{\partial H}{\partial \omega} > 0 \tag{5.80}$$

where the asterisk indicates the complex conjugate. If the imaginary part of a vector is positive then the angle of the vector lies between $0°$ and $180°$. Since the conjugate possesses a negative angle Equation 5.80 requires that the angle of the derivative with regard to ω minus the angle of the derivative with regard to A must be between $0°$ and $180°$. This is shown graphically in Figure 5.14(a). For a stable limit cycle the derivative with regard to A must lie to the right of the dashed lines when the derivative with regard to ω has the direction shown.

If the variables are separable then

$$H(A, \omega) = N(A)G(j\omega) \tag{5.81}$$

and at the equilibrium point where $1 + H = 0$ it is true that

$$G(j\omega) = - \frac{1}{N(A)} \tag{5.82}$$

The partial derivatives of H can now be expressed as

$$\frac{\partial H}{\partial A} = G \frac{\partial N}{\partial A} = -N \frac{\partial}{\partial A} \left(-\frac{1}{N} \right)$$
$$\frac{\partial H}{\partial \omega} = N \frac{\partial G}{\partial \omega} \tag{5.83}$$

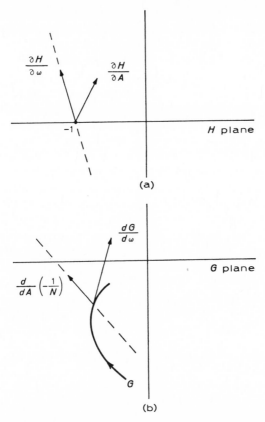

Figure 5.14   Geometric relationships for a stable
limit cycle.

Due to the factors $-N$ and N the sense of the stability criterion must be
reversed and a stable limit cycle is indicated by the relationships shown in
part (b) of Figure 5.14. Each derivative is tangent to its curve at the point of
intersection and their length and direction indicate the change in each
function as A and ω increase.

   Goldfarb has suggested an alternate method for determining the stability
of a limit cycle. A translation of one of his reports is listed in the bibliography.
After determining the values of A and ω which exist at the equilibrium pooit
the locus of $N(A + \delta A)G(j\omega)$ is constructed in the neighborhood of the
critical point at $(-1, 0)$. If this locus encircles the critical point when δA is
positive then the oscillations are unstable since δA would increase further.
It is left as an interesting exercise to compare the methods of Loeb and
Goldfarb.

It is possible for the curves to be related as shown in Figure 5.15. This shows four possible limit cycles. Application of the Loeb criterion shows that A and C are stable while B and D are unstable. If the system should be oscillating in mode C then under a slight disturbance it will return to this mode. If the amplitude is increased sufficiently it will pass through the unstable mode B and settle at A. If the amplitude is decreased sufficiently the system will pass through the unstable mode D and the oscillation will die completely. This property of possessing several stable modes with intermediate unstable modes is entirely unique with nonlinear networks and systems. It is known as a *jump phenomenon*. The particular case of jump resonance has been explained rather satisfactorily using the technique of the *dual-input describing function*. Several reports are listed as references for this technique. It requires extensive computation and will not be discussed here.

The closed-loop frequency response of a linearized system is obviously a function of the input amplitude since the nonlinearity is represented by a describing function. Therefore there is no single characteristic but instead a family of curves with amplitude as the parameter. After the plots of G and $-1/N$ have been made it is fortunately very easy to derive the information for any one of these curves. The technique depends on writing the closed-loop

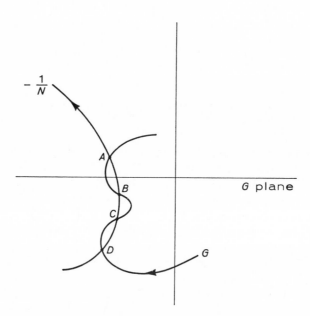

Figure 5.15   Alternate stable and unstable limit cycles.

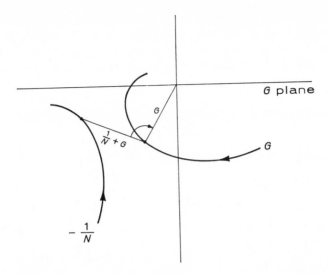

Figure 5.16 Closed-loop frequency response of a linearized system.

transfer function as

$$\frac{NG}{1+NG} = \frac{G}{\dfrac{1}{N}+G} \tag{5.84}$$

and observing that the numerator and denominator are both available graphically as shown in Figure 5.16. The phase angle is also indicated in the proper sense. It is necessary to select a value for A and locate the corresponding point on the $-1/N$ curve. This point remains stationary while measurements are made to points on the plot of G corresponding to various values of ω. The ratios of the lengths are computed and the angles measured to obtain the closed-loop amplitude and phase response.

There is good reason to doubt the value of the knowledge of the sinusoidal response of a system. Such inputs are almost never used except for tests. Typical inputs vary in time in such a manner that statistical methods must be used to analyze them. But statistical methods can give only a statistical answer. As a result it is difficult to predict the time dependent response and the amount of error or misbehavior which would be present. Nevertheless, the sinusoidal response gives an intuitive form of information about the system and provides a convenient comparison between systems. For these reasons it will probably continue to be used.

In concluding this section on linearized systems it is probably wise to state once more that all results and conclusions depend on the validity of the approximation using the describing function. If the linear portion of the system acts as a low-pass filter or if the nonlinearity generates harmonics at low amplitudes then the approximation can be expected to give good results. Otherwise the opposite may be true. The methods described in the following chapter treat the nonlinearity without linearization and, though limited in scope of applicability, provide mathematical rigor.

REFERENCES

1. D. D. Šiljak, *Nonlinear Systems*, Wiley, New York, 1969.
2. D. Graham and D. Ruer, *Analysis of Nonlinear Control Systems*, Wiley, New York, 1961.
3. H. Nyquist, Regeneration Theory, *Bell System Technical Journal*, Volume 11, 1932, pp. 126–247.
4. N. Kryloff and N. Bogoliuboff, *Introduction to Nonlinear Mechanics*, Princeton University Press, 1947.
5. R. Bass, Mathematical Legitimacy of Equivalent Linearization by Describing Functions, *Automatic and Remote Control*, Butterworth's, London, 1961.
6. J. L. Douce, A Note on the Evaluation of the Response of a Nonlinear Element to Sinusoidal and Random Signals, *Proceedings of the IEE*, Volume 105C, 1958, pp. 88–92.
7. T. Stout, Block Diagram Transformations for Systems with One Nonlinear Element, *Transactions of the AIEE*, Number 25, 1956, pp. 130–141.
8. J. M. Loeb, Recent Advances in Nonlinear Servo Theory, *Frequency Response*, Macmillan, New York, 1956.
9. L. Goldfarb, On Some Nonlinear Phenomena in Regulatory Systems, *Frequency Response*, Macmillan, New York, 1956.
10. J. West, J. Douce, and R. Livesley, The Dual-Input Describing Function and its Use in the Analysis of Nonlinear Feedback Systems, *Proceedings of the IEE*, Volume 103B, 1956, pp. 463–474.
11. R. M. Huey, O. Pawloff, and T. Glucharoff, Extension of the Dual-Input Describing-Function Technique to Systems Containing Reactive Nonlinearity, *Proceedings of the IEE*, Volume 107C, 1960, pp. 334–341.

6

THE PHASE PLANE

The methods of the last chapter incorporate a representation of the nonlinear aspect of the system in a manner which permits the application of standard linear analysis with appropriate extensions. The expressions are approximate and there is always doubt as to the accuracy of the results. This chapter discusses a method of considerable generality in which all nonlinear characteristics are treated without modification or linearization. There is therefore no intrinsic limitation on accuracy. However, the technique is usually graphical and almost always incremental or iterative.

In its simplest form the method applies only to second order systems wherein there is no explicit time dependency. Methods are presented later in the chapter whereby both of these restrictions can be removed although the technique becomes considerably more complicated. The chapter includes a survey of a discipline traditionally called *nonlinear mechanics* since this method of analysis relies heavily on the use of the phase plane and the results are of considerable interest.

To apply the phase plane method it is necessary and sufficient that the basic differential equation can be written in the general form

$$\ddot{x} - f(x, \dot{x}) = 0 \qquad (6.1)$$

where any nonlinear characteristics are expressed in $f(x, \dot{x})$. The substitution

$$\ddot{x} = \frac{d\dot{x}}{dt} = \frac{dx}{dt}\frac{d\dot{x}}{dx} = \dot{x}\frac{d\dot{x}}{dx} \qquad (6.2)$$

reduces the apparent order and the equation assumes the form

$$\frac{d\dot{x}}{dx} = \frac{f(x, \dot{x})}{\dot{x}} \qquad (6.3)$$

Although time remains the true independent variable it has disappeared from the basic equation. The new equation has only the two variables x and

$\dot{x}$. It is now possible to consider a solution of the form

$$\dot{x} = \dot{x}(x) \tag{6.4}$$

which suggests plotting $\dot{x}$ against x. The plane containing the x and $\dot{x}$ axes is called the *phase plane*. A solution $x(t)$ of the original second order equation appears as a curve in the phase plane. Such a curve is called a *trajectory*. Figure 6.1 shows general forms of trajectories. All of these would normally not occur under the same conditions in the same system. Since $\dot{x}$ determines the movement of x in time it is usually true that when $\dot{x}$ is positive the trajectory moves only to the right and when $\dot{x}$ is negative it moves only to the left. Also since $\dot{x} = 0$ along the x axis the trajectories usually cross the x axis perpendicularly. None of these are sufficient conditions, however, and Figure 6.19 shows trajectories which violate all of these rules.

If $f(x, \dot{x})$ is a single-valued function of x and $\dot{x}$ then Equation 6.3 defines the slope of a trajectory uniquely at each point. Under this quite common condition it is impossible for trajectories to cross. If the condition does not exist then it is necessary to divide the phase plane into layers at branch points and treat each possibility separately. Since infinite acceleration is impossible the trajectories must be continuous curves.

If the right side of Equation 6.3 is equated to a constant M giving

$$\frac{f(x, \dot{x})}{\dot{x}} = M \tag{6.5}$$

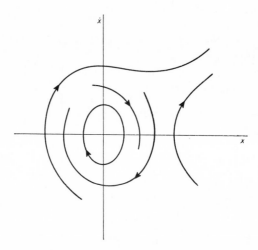

Figure 6.1 The general form of phase plane trajectories.

and this expression written as

$$f(x, \dot{x}) - M\dot{x} = 0 \tag{6.6}$$

the result is an equation which defines the locus of points in the phase plane wherein the trajectories have the slope M. Such a locus is called an *isocline*. If Equation 6.6 can be solved for either x or $\dot{x}$ it can be placed in one of the forms

$$x = x(M, \dot{x})$$
$$\dot{x} = \dot{x}(M, x) \tag{6.7}$$

Using M as a parameter it is possible to draw a set of isoclines throughout the phase plane. These may then be used to determine trajectories graphically. The trajectory starts at the point corresponding to the initial values of x and $\dot{x}$. It then follows the path through the phase plane which is consistent with the requirements of the isoclines.

Two methods for improving the accuracy of the geometrical construction are shown in Figure 6.2. In the method shown in part (a) the lines from point P_1 have slopes M_1 and M_2 which are indicated by the short line segments crossing each isocline. They are extended until they meet the isocline indicating slope M_2. The point P_2 is midway between the intersections. The process is then continued. The trajectory is plotted as a smooth curve through the points $P_1, P_2, \ldots$ which are thus determined. In this method a mean value is used for the slope between isoclines.

In the alternative shown in part (b) of Figure 6.2 the slope between Q and R has the value M_1 and the slope between R and S has the value M_2 as required by their respective isoclines. There is no specific criterion for determining the location of the point R other than the general consideration that it should lie approximately midway between the isoclines. The trajectory is again plotted as a smooth curve through the points $P_1, P_2, \ldots$ which are thus determined. In this method the proper value for the slope is used on both sides of the isocline. Neither method is always superior and the actual geometry which is involved should indicate the preferred method.

Before closing this general discussion of isoclines a few additional properties should be mentioned. Letting $M = 0$ in Equation 6.6 generates the equation for a special curve known as the *zero slope isocline*. This curve has the obvious geometrical significance that it describes the locus where trajectories are horizontal and is therefore analogous to the x axis where trajectories usually are vertical. It also corresponds to the situation where $f(x, \dot{x}) = 0$. Returning to the original differential equation it is observed that this implies zero acceleration and zero force. In physical systems it is frequently useful to know the conditions under which such a situation exists.

In all realizable physical systems there are values of x and $\dot{x}$ which make the right hand side of Equation 6.3 the indeterminate ratio of zero divided by

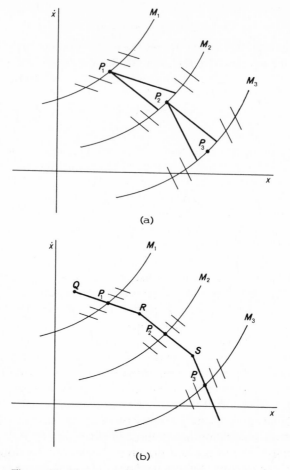

Figure 6.2 Two methods for using isoclines to determine trajectories.

zero. This almost always occurs along the x axis where $\dot{x} = 0$ but if $f(x, \dot{x})$ has a form such as

$$f(x, \dot{x}) = \dot{x}\frac{(x - x_0)^m + (\dot{x} - \dot{x}_0)^n}{(x - x_0)^r + (\dot{x} - \dot{x}_0)^s} \tag{6.8}$$

then these unique points can be translated to any desired point in the phase plane. The only way for the slope to be indeterminate is for isoclines indicating all values of slope to pass through these unusual points. This peculiar type of point on the phase plane is called a *singular point*. When defined as a unique condition in a set of nonlinear differential equations such a point is called a *critical point* and this terminology is frequently encountered. However, in the present context such points will be known only as singular points. The fact that no unique motion is defined at these points means that they represent conditions of static equilibrium. This may be stable or unstable depending on the behavior of the system in the region around the point. It is to be expected that trajectories behave in a special manner relative to these points. This is entirely true. An extensive discussion of the properties of trajectories and other curves around singular points is presented later in this chapter.

THE TIME INTERVAL

Most methods for determining the amount of time required to traverse a section of a trajectory depend on the fundamental relation

$$T = \int_0^T dt = \int_{x_0}^x \frac{dx}{\dot{x}} \tag{6.9}$$

and the methods differ only in the manner in which the quantities are evaluated. The integration is usually performed as the sum of finite increments which are generated as the trajectory is developed. No method is always superior to the others. The relative convenience and accuracy depend on the location and orientation of the trajectory. Fortunately, different methods can be used in different portions of the trajectory and the results combined to get the total time interval.

The most direct method is to make an additional plot of $1/\dot{x}$ as the plot of $\dot{x}$ is developed. The time interval is the area under this curve. This is shown in Figure 6.3. The area can be found in any desired manner but it is conveniently calculated as the sum of incremental values of dx multiplied by appropriate values of $1/\dot{x}$. Mean values will improve accuracy. The method is difficult to apply when the trajectory crosses the x axis since $1/\dot{x}$ runs to infinity.

An alternate method which frequently provides improved accuracy in this region depends on the relations

$$dt = \frac{dx}{\dot{x}} = \frac{d\dot{x}}{f(x, \dot{x})} \tag{6.10}$$

In this case the function $1/f$ must be calculated as the trajectory is developed. The area to the $\dot{x}$ axis must be calculated and this involves the incremental values of $d\dot{x}$. The situation is shown in Figure 6.4.

A method which requires no additional calculation is shown in Figure 6.5. The line segment PQ is perpendicular to the incremental segment ds. The technique depends on the relations

$$dt = \frac{dx}{\dot{x}} = \frac{ds}{r} = \tan d\alpha \approx d\alpha \tag{6.11}$$

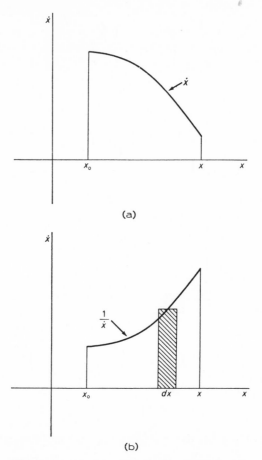

Figure 6.3 Finding the time interval. (a) The original trajectory and (b) the inverse plot.

Proportionality between corresponding sides of similar triangles verifies the construction. The size of the incremental angle in radians equals the incremental time interval in seconds. The method is difficult to apply near the x axis where the slope of the trajectory may approach infinity and the angle α may become so small that the determination of the point P may be inaccurate.

This difficulty can be alleviated by observing another interesting feature of the construction. Due to similarity of the triangles it is true that

$$\frac{d\dot{x}}{dx} = \frac{\Delta x}{\dot{x}} = \frac{\ddot{x}}{\dot{x}}$$

$$\Delta x = \ddot{x} = f(x, \dot{x})$$

(6.12)

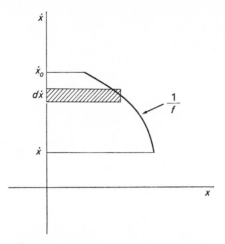

Figure 6.4   Alternate method for finding the time interval.

so the projection of PQ indicates the acceleration. The indicated sense is positive. This relationship is quite general. It applies at any time to any trajectory. This principle can be used to improve the above construction by using a calculated value of $f(x, \dot{x})$ to locate the point Q when the geometrical construction becomes inaccurate.

Frequently it is desired to tabulate information about the trajectory at points which are equally spaced in time. This can be done by the construction

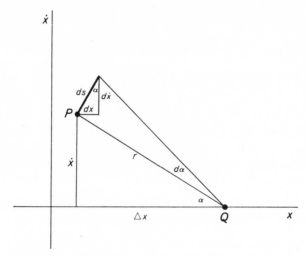

Figure 6.5   Finding the acceleration and the time interval.

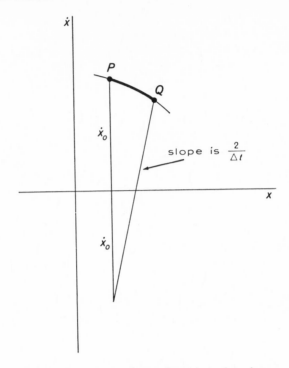

Figure 6.6   A construction to determine points along the trajectory equally spaced in time.

shown in Figure 6.6. Between points P and Q the mean value of velocity is approximately given by

$$\dot{x} = \dot{x}_0 + \frac{\Delta \dot{x}}{2} = \frac{\Delta x}{\Delta t} \tag{6.13}$$

and this can be written as

$$\Delta \dot{x} = \left(\frac{2}{\Delta t}\right) \Delta x - 2\dot{x}_0 \tag{6.14}$$

For increments at P this is the equation of a straight line with intercept $-2\dot{x}_0$ and slope $2/\Delta t$. The intersection of this line with the trajectory determines Q. Maintaining constant slope for this additional straight line as the construction progresses gives points along the trajectory which are equally spaced in time.

FORMS OF TRAJECTORIES

To discuss the many forms which the trajectories can assume it is most instructive to consider the various forms which are present in linear systems and then to project these results into the nonlinear domain. To simplify the algebra it is assumed that the linear system is represented by the equation

$$\ddot{x} + 2p\dot{x} + qx = 0 \tag{6.15}$$

In the phase plane representation this becomes

$$\frac{d\dot{x}}{dx} = \frac{-2p\dot{x} - qx}{\dot{x}} \tag{6.16}$$

The characteristic equation of the original differential equation is quadratic and in general the roots are given by

$$\begin{aligned}
\lambda_1 &= -p - \sqrt{p^2 - q} \\
\lambda_2 &= -p + \sqrt{p^2 - q}
\end{aligned} \tag{6.17}$$

In a physically realizable system the coefficients p and q must be real. However, they can assume any combination of positive and negative values. As a consequence the above roots can be (a) complex, (b) purely imaginary, (c) real and of the same sign, (d) real and of opposite sign, and (e) equal. Each of these cases produces a different form for the trajectory and each must be treated separately.

In the first case to be considered it is assumed that $q > p^2$ and the roots are necessarily complex. The substitution

$$\omega = \sqrt{q - p^2} \tag{6.18}$$

permits the solution for x and $\dot{x}$ to be written as

$$\begin{aligned}
x &= Ae^{-pt} \sin(\omega t + \Phi) \\
\dot{x} &= -pAe^{-pt} \sin(\omega t + \Phi) + \omega Ae^{-pt} \cos(\omega t + \Phi)
\end{aligned} \tag{6.19}$$

It is possible to consider t as a parameter and to plot corresponding values of x and $\dot{x}$ to determine the trajectory. This is tedious. A more clever technique is to introduce new variables which are chosen so as to simplify the expressions. Such a transformation is

$$\begin{aligned}
u &= \omega x = \omega Ae^{-pt} \sin(\omega t + \Phi) \\
v &= px + \dot{x} = \omega Ae^{-pt} \cos(\omega t + \Phi)
\end{aligned} \tag{6.20}$$

If u and v are plotted in a rectangular coordinate system then it is possible to define

$$\begin{aligned}
\rho^2 &= u^2 + v^2 = \omega^2 A^2 e^{-2pt} \\
\theta &= 90° - (\omega t + \Phi)
\end{aligned} \tag{6.21}$$

and these are the equations of a logarithmic spiral with t as the parameter. If $p > 0$ the motion is toward the origin as t increases. Since the origin in the

u, v plane is also the origin in the $x, \dot{x}$ plane this indicates an oscillatory motion which converges to the singular point in the $x, \dot{x}$ plane. Conversely, if $p < 0$ the motion is divergent from the origin and this indicates an oscillation of increasing amplitude.

To determine the appearance of the spiral in the $x, \dot{x}$ plane it is necessary to examine a few properties of the transformation which was used. The equation $u = \omega x$ which is true for all v and all $\dot{x}$ shows that there is everywhere a constant relation between horizontal distances in the two planes. When $u = x = 0$ then $\dot{x} = v$ so the intercepts along the vertical axes are equal. If v remains constant in the equation $\dot{x} = v - px$ the result is the equation for straight lines with slope $-p$. Therefore the transformation from the u, v plane to the $x, \dot{x}$ plane involves (a) a horizontal scale factor of ω, (b) a vertical scale factor of unity, and (c) a rotation of horizontal lines in the u, v plane to lines with a slope of $-p$ in the $x, \dot{x}$ plane.

It is also informative and useful to know the slope of trajectories as they cross the axes. Reference to Equation 6.16 shows that along the x axis where $\dot{x} = 0$ the slope is infinite and along the $\dot{x}$ axis where $x = 0$ the slope of every trajectory is $-2p$. Furthermore, the zero slope isocline is given by the relation

$$\dot{x} = -\frac{q}{2p} x \tag{6.22}$$

These values are not dependent on this particular transformation and apply to all trajectories of the linear system.

In the u, v plane the time derivatives are

$$\frac{du}{dt} = \omega \frac{dx}{dt}$$
$$\frac{dv}{dt} = p \frac{dx}{dt} + \frac{d\dot{x}}{dt} \tag{6.23}$$

so the space derivative is given by

$$\frac{dv}{d_n} = \frac{p}{\omega} + \frac{1}{\omega} \frac{d\dot{x}}{dx} \tag{6.24}$$

Along the v axis where $u = 0$ and $x = 0$ the slope in the $x, \dot{x}$ plane is $-2p$ so in the u, v plane it is

$$\frac{dv}{du} = \frac{p}{\omega} - \frac{2p}{\omega} = -\frac{p}{\omega} \tag{6.25}$$

Along the u axis where $v = 0$ and $\dot{x} = -px$ appropriate substitutions show that

$$\frac{dv}{du} = \frac{p}{\omega} + \frac{q - 2p^2}{p\omega} = \frac{\omega}{p} \tag{6.26}$$

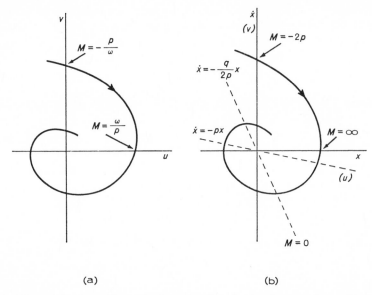

$$(a) \qquad\qquad\qquad (b)$$

Figure 6.7 Comparison of trajectories when the singular point is a stable focus.

A combination of all the above considerations shows that the trajectories in both planes will have the general appearance shown in Figure 6.7. When the trajectories have this form the singular point is called a *focus*. The figure shows a stable focus wherein the u axis is rotated clockwise and the spiral moves inward. This represents a damped vibration.

The second type of trajectory to be considered is a special case of the first. If $q > p^2$ and also $p = 0$ then the roots of the characteristic equation are purely imaginary. In this case the appropriate transformation is

$$u = \omega x = \omega A \sin (\omega t + \Phi)$$
$$v = \dot{x} = \omega A \cos (\omega t + \Phi) \tag{6.27}$$

and these are the parametric equations of a circle with radius ωA. Since $p = 0$ there is no rotation of the u axis and the circle is transformed into an ellipse in the x, $\dot{x}$ plane. This form of vorticose motion corresponds to a continuous oscillation. The trajectories are shown in Figure 6.8. In this case the singular point is called a *center*.

In the third case to be considered it is assumed that $q < p^2$ and the roots of the original characteristic equation are therefore real and distinct. The displacement and velocity are expressed as

$$x = Ae^{\lambda_1 t} + Be^{\lambda_2 t}$$
$$\dot{x} = \lambda_1 Ae^{\lambda_1 t} + \lambda_2 Be^{\lambda_2 t} \tag{6.28}$$

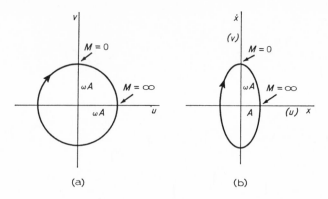

Figure 6.8   Steady oscillation around a center.

The separate solutions can be decoupled by the transformation

$$u = \dot{x} - \lambda_2 x = (\lambda_1 - \lambda_2)Ae^{\lambda_1 t}$$
$$v = \dot{x} - \lambda_1 x = (\lambda_2 - \lambda_1)Be^{\lambda_2 t} \tag{6.29}$$

which can be written in the simpler forms

$$u = u_0 e^{\lambda_1 t}$$
$$v = v_0 e^{\lambda_2 t} \tag{6.30}$$

which represent simple exponential solutions. The form of the solution paths in the u, v plane depends on the values of λ_1 and λ_2. It is first assumed that $q > 0$ and $p < 0$ so

$$\lambda_2 < \lambda_1 < 0 \tag{6.31}$$

Since both roots are negative both u and v approach the origin as time passes. Also if the roots have the assumed relative values then the ratio of v to u approaches zero as time becomes infinite so the solution paths are tangent to the u axis at the origin. They therefore have the general appearance shown in part (a) of Figure 6.9. To transform these solutions to the x, $\dot{x}$ plane it useful to observe a few properties of this particular transformation. Points along the u axis where $v = 0$ are transformed to the line $\dot{x} = \lambda_1 x$ in the $x,\dot{x}$ plane. Similarly, the v axis becomes the line $\dot{x} = \lambda^2 x$. The transformation therefore introduces a rotation of both axes. Letting u and v have the constant values u_0 and v_0 gives the equations

$$\dot{x} = u_0 + \lambda_2 x$$
$$\dot{x} = v_0 + \lambda_1 x \tag{6.32}$$

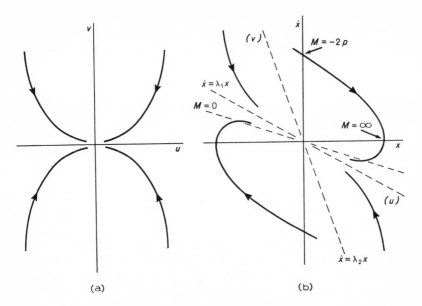

Figure 6.9   Trajectories for a stable node.

which are the equations of straight lines with intercepts of u_0 and v_0 and with slopes of λ_2 and λ_1. These lines are parallel to the transformed u and v axes. From the above it is concluded that this more general transformation (a) retains the origin, (b) rotates both axes, (c) preserves intercepts along rectangular systems, and (d) maintains proportionality in both directions. With the help of these rules it is not difficult to understand that the trajectories in the x, $\dot{x}$ plane have the forms shown in part (b) of Figure 6.9. Since it is assumed that $p > 0$ the obvious inequality

$$\left(p - \sqrt{p^2 - q}\right)^2 > 0 \tag{6.33}$$

leads to the result

$$|\lambda_1| - \frac{q}{2p} > 0 \tag{6.34}$$

and the zero slope isocline must be between the u axis and the x axis. This is consistent with the geometrical requirements on the curves. This type of trajectory represents aperiodic motion to an equilibrium point. This type of singular point is called a *node*. If both roots are positive the motion is divergent and the node is unstable.

In the fourth case to be considered it is assumed that $q < 0$ and regardless of the value of p the roots are real and of opposite sign. Assuming further that $\lambda_1 > 0$ and $\lambda_2 < 0$ then the trajectories are asymptotic to the u axis and have the appearance shown in part (a) of Figure 6.10. In this case the transformation to the x, $\dot{x}$ plane is more nearly a pure rotation since the u axis has a positive slope. Under the assumptions for q and p the obvious inequality

$$\left(\sqrt{p} - \sqrt{p - \frac{q}{p}}\right)^2 > 0 \tag{6.35}$$

leads to the result

$$\left|\frac{q}{2p}\right| - \lambda_1 > 0 \tag{6.36}$$

and the zero slope isocline must lie between the u axis and the $\dot{x}$ axis. Part (b) of Figure 6.10 shows the appearance of the trajectories in the x, $\dot{x}$ plane. It is quite clear that the motion consistently avoids the origin which must therefore be a point of unstable equilibrium. This type of singular point is called a *saddle point*. The name is derived from the shape of a surface which occurs in a similar situation of higher dimension.

In the fifth and last case to be considered it is assumed that $q = p^2$. There is a double root and $\lambda = -p$. The general solutions in this case have the form

$$x = (A + Bt)e^{\lambda t}$$
$$\dot{x} = \lambda(A + Bt)e^{\lambda t} + Be^{\lambda t} \tag{6.37}$$

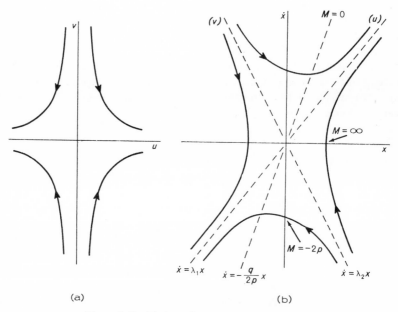

(a)   (b)

Figure 6.10 Trajectories around a saddle point.

These equations, unlike the others, cannot be completely decoupled. A partial separation is accomplished by the transformation

$$u = x = (A + Bt)e^{\lambda t}$$
$$v = \dot{x} - \lambda x = Be^{\lambda t}$$

(6.38)

Immediately it is observed that the ratio of u to v becomes infinite as time passes so the trajectories are asymptotic to the u axis. Since the time derivatives are given by

$$\frac{du}{dt} = \lambda u + v$$
$$\frac{dv}{dt} = \lambda v$$

(6.39)

the space derivative is given by

$$\frac{dv}{du} = \frac{\lambda v}{\lambda u + v}$$

(6.40)

which is infinite along the straight line $v = -\lambda u$ and which has the value λ along the v axis where $u = 0$. The u axis lies on the line $\dot{x} = \lambda x$. Assuming $p > 0$ the inequality

$$\frac{q}{2p} = \frac{p}{2} < p$$

(6.41)

shows that the zero slope isocline lies between the u axis and the x axis. Also, if $p > 0$ then $\lambda < 0$ and both u and v approach zero as time becomes infinite so the trajectory approaches the origin. A combination of the above characteristics shows that the trajectories in each plane have the appearance shown in Figure 6.11. These resemble the trajectories in Figure 6.9 and this type of singular point is also called a node although the mathematical form of the solution is quite different.

This completes the discussion of the trajectories which occur in a linear system. It is somewhat astonishing that so many forms exist for such a simple system as that described by Equation 6.15. It is to be expected that nonlinear systems present an even greater variety of forms and this is true.

Nonlinear processes usually behave in nearly a linear fashion in a sufficiently small region around the operating point. This suggests the possibility that in a sufficiently small region around a singular point the trajectories of a nonlinear system appear essentially the same as those of its linear counterpart. Oddly, this is not always true. The proper relationship is stated in the following theorem due to Liapunov which has a central position in phase plane theory.

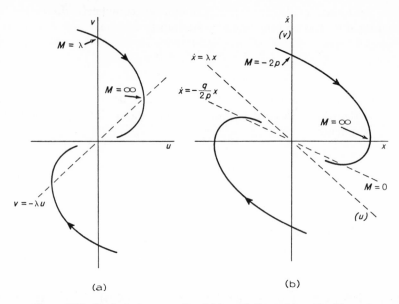

Figure 6.11   Trajectories around a stable node formed by a double root.

Theorem 6.1   *If the nonlinear characteristic of a system is analytic and contains linear terms then its trajectories and stability in a small region around a singular point are the same as those in its linear counterpart except that if the linear system has a center then the nonlinear system may have a center or a focus and the focus may be stable or unstable.*

The proof of this theorem is necessarily lengthy and complex since it involves the behavior of a nonlinear system around the various types of singular points found in linear systems. The method of proof here establishes a simple geometric relationship between the trajectories for aperiodic motion around nodal and saddle points and then uses a more powerful method for comparing vorticose motion. The nonlinear system is represented by

$$\frac{d\dot{x}}{dx} = \frac{Ax + B\dot{x} + \phi(x, \dot{x})}{Cx + D\dot{x} + \psi(x, \dot{x})} = \frac{R(x, \dot{x}) + \phi(x, \dot{x})}{S(x, \dot{x}) + \psi(x, \dot{x})} \qquad (6.42)$$

and the corresponding linear system is expressed by

$$\frac{d\dot{x}}{dx} = \frac{Ax + B\dot{x}}{Cx + D\dot{x}} = \frac{R(x, \dot{x})}{S(x, \dot{x})} \qquad (6.43)$$

These expressions are somewhat more general than those which would be derived directly from Equation 6.16 but they introduce little additional complexity and they are mathematically convenient. Each of the above derivatives can be considered as a vector which is everywhere tangent to its trajectory. In each case, the x and $\dot{x}$ components of the vector are the denominator and numerator respectively. Since the phase plane is an orthogonal system the techniques of ordinary vector analysis may be employed. In particular, the cosine of the angle between the vectors is equal to the dot product divided by the product of the lengths. When this principle is applied to the above vectors the result is

$$\cos \theta = \frac{R(R + \phi) + S(S + \psi)}{(R^2 + S^2)^{\frac{1}{2}}[(R + \phi)^2 + (S + \psi)^2]^{\frac{1}{2}}}$$

$$= \frac{1 + \dfrac{R\phi + S\psi}{R^2 + S^2}}{\left\{ 1 + \left[\dfrac{2(R\phi + S\psi)}{R^2 + S^2} \right] + \dfrac{\phi^2 + \psi^2}{R^2 + S^2} \right\}^{\frac{1}{2}}} \tag{6.44}$$

Considering the function

$$R^2 + S^2 = (Ax + B\dot{x})^2 + (Cx + D\dot{x})^2 \tag{6.45}$$

it is first apparent that it can be only positive or zero. Assuming that the singular point for the linear system is at the origin then the function can be zero only at that point if the situation is not trivial. It is therefore possible to select a path such as a circle around the origin along which and within which the function is never zero. Along the path the function has a minimum which shall be called m. Then for all points in the phase plane it is true that

$$R^2 + S^2 \geqslant m\{|x| + |\dot{x}|\}^2 \tag{6.46}$$

This inequality can be established by considering points along a ray defined by $\dot{x} = kx$. The substitution eliminates one variable and varying values of k sweep the entire plane. It is also quite obviously true that

$$\sqrt{R^2 + S^2} \geqslant R, \qquad \sqrt{R^2 + S^2} \geqslant S \tag{6.47}$$

Applying these inequalities to Equation 6.44 gives the results

$$\left| \frac{R\phi + S\psi}{R^2 + S^2} \right| \leqslant \frac{|\phi| + |\psi|}{R^2 + S^2} \leqslant \frac{1}{m^{\frac{1}{2}}} \frac{|\phi| + |\psi|}{|x| + |\dot{x}|}$$

$$\left| \frac{\phi^2 + \psi^2}{R^2 + S^2} \right| \leqslant \frac{\{|\phi| + |\psi|\}^2}{R^2 + S^2} \leqslant \frac{1}{m}\left(\frac{|\phi| + |\psi|}{|x| + |\dot{x}|} \right)^2 \tag{6.48}$$

So $\cos \theta$ can be made as nearly unity or θ can be made as small as desired *uniformly* in a region around the origin if

$$\lim_{|x| + |\dot{x}| \to 0} \frac{|\phi| + |\psi|}{|x| + |\dot{x}|} = 0 \tag{6.49}$$

This represents a requirement on the nature of the nonlinearity for the theorem to be valid. If, however, the nonlinear functions are analytic and may be expanded into series or represented as polynomials by interpolation then ϕ and ψ contain only higher order terms and the above requirement is certainly met.

It may appear that the above constitutes a satisfactory proof of the theorem. This is not true. Vorticose motion is not suitably defined by the above geometric property. In particular it is possible for a spiral about a focus to intersect oscillatory trajectories about a center with vanishing angles as it approaches the origin. To prove the theorem for vortical motion it is unfortunately necessary to examine the exact nature of the trajectories which may exist in the nonlinear system when a focus or a center is present in the linear system. To do this, recourse must be taken to mathematical methods of greater power and generality.

The technique consists of the use of linear transformations such as were employed in the discussion of the trajectories present in linear systems. The method is developed for a linear system and then extended to nonlinear systems. The second order linear system is represented by the equations

$$\frac{d\dot{x}}{dt} = Ax + B\dot{x}$$

$$\frac{dx}{dt} = Cx + D\dot{x}$$

(6.50)

If the first is divided by the second the result is Equation 6.43. Again the expression is far more general than the equation which would be derived directly from Equation 6.16. In particular C should equal zero and D should be unity. But the excess generality introduces very little additional complexity, it is mathematically satisfying, and in the nonlinear case it may have significance.

It has been found that with proper selection of the constants a transformation such as

$$u = ax + b\dot{x}$$

$$v = cx + d\dot{x}$$

(6.51)

produces a set of equations in the form

$$\dot{u} = \lambda u$$

$$\dot{v} = \mu v$$

(6.52)

The constants are found by differentiating Equations 6.51 and substituting Equations 6.50 to give

$$\dot{u} = a\dot{x} + b\ddot{x} = a(Cx + D\dot{x}) + b(Ax + B\dot{x})$$

$$\dot{v} = c\dot{x} + d\ddot{x} = c(Cx + D\dot{x}) + d(Ax + B\dot{x})$$

(6.53)

Substituting Equations 6.51 into Equations 6.52 and comparing the coefficients of x and $\dot{x}$ in the last two sets of equations gives the relations

$$\lambda = \frac{Ca + Ab}{a} = \frac{Bb + Da}{b}$$

$$\mu = \frac{Cc + Ad}{c} = \frac{Bd + Dc}{d} \tag{6.54}$$

Simultaneous solution of these equations provides the separate relationships

$$a = \left(\frac{\lambda - B}{D}\right)b$$

$$b = \left(\frac{\lambda - C}{A}\right)a$$

$$c = \left(\frac{\mu - B}{D}\right)d \tag{6.55}$$

$$d = \left(\frac{\mu - C}{A}\right)c$$

Combining corresponding pairs gives

$$\lambda^2 - (B + C)\lambda - (AD - BC) = 0$$
$$\mu^2 - (B + C)\mu - (AD - BC) = 0 \tag{6.56}$$

so the constants λ and μ are roots of the characteristic equation resulting from simultaneous solution of Equations 6.50. Since there are two basic restrictions on the four constants one from each of the sets a, b and c, d may be selected arbitrarily. Any set of four constants selected in this manner reduce Equations 6.50 to Equations 6.52.

For the nonlinear equations

$$\frac{d\dot{x}}{dt} = P(x, \dot{x}) = Ax + B\dot{x} + \phi(x, \dot{x})$$

$$\frac{dx}{dt} = Q(x, \dot{x}) = Cx + D\dot{x} + \psi(x, \dot{x}) \tag{6.57}$$

the above transformation yields the set

$$\dot{u} = \lambda u + b\phi'(u, v) + a\psi'(u, v)$$
$$\dot{v} = \mu v + d\phi'(u, v) + c\psi'(u, v) \tag{6.58}$$

where ϕ' and ψ' contain only higher order and cross-product terms in u and v. They are the direct result of the substitution of u and v for x and $\dot{x}$. Since only vorticose motion remains to be discussed the only case to be considered is that wherein the roots are complex conjugates and

$$\lambda = \alpha + j\omega$$
$$\mu = \alpha - j\omega$$

(6.59)

This suggests the fact that u and v are also complex conjugates and can be expressed as

$$u = U + jV$$
$$v = U - jV$$

(6.60)

A substitution in Equation 6.58 gives

$$\dot{u} = \dot{U} + j\dot{V} = \lambda(U + jV) + F(U, V)$$
$$\dot{v} = \dot{U} - j\dot{V} = \mu(U - jV) + F^*(U, V)$$

(6.61)

where F^* is the conjugate of F and both contain only higher order terms. Adding gives

$$2\dot{U} = (\lambda + \mu)U + j(\lambda - \mu)V + (F + F^*)$$

(6.62)

and subtracting gives

$$2j\dot{V} = (\lambda - \mu)U + j(\lambda + \mu)V + (F - F^*)$$

(6.63)

These are equivalent to

$$\dot{U} = \alpha U - \omega V + F_u$$
$$\dot{V} = \omega U + \alpha V + F_v$$

(6.64)

where all terms are real. Defining a radial distance in the U, V plane as

$$R^2 = U^2 + V^2$$

(6.65)

provides the relation

$$\frac{dR^2}{dt} = 2U\dot{U} + 2V\dot{V}$$

(6.66)

This expression can be formed by multiplying Equations 6.64 by $2U$ and $2V$ respectively and then adding. After a slight simplification this gives

$$\frac{dR^2}{dt} = 2\alpha R + 2UF_u + 2VF_v$$

(6.67)

Changing to polar coordinates where

$$U = R\cos\theta, \qquad V = R\sin\theta$$

(6.68)

and observing that

$$\frac{dR^2}{dt} = 2R\frac{dR}{dt}$$

(6.69)

permit a reduction of Equation 6.67 to

$$\frac{dR}{dt} = \alpha R + G(\theta, R) \tag{6.70}$$

where G contains only higher order terms in R and no constant terms. It is therefore possible to remove R as a factor and this gives

$$\frac{dR}{R} = [\alpha + H(\theta, R)] \, dt \tag{6.71}$$

where H contains R at least to the first degree. Since H vanishes at least linearly with R for any θ it is possible to select a circle with its center at the origin wherein $|H|$ is everywhere less than an arbitrary, positive ϵ. Integration of the above equation starting at a point within the circle produces relations which may be written in the form

$$\log \frac{R}{R_0} \geqslant (\alpha - \epsilon)t, \quad \alpha > 0$$
$$\log \frac{R}{R_0} \leqslant (\alpha + \epsilon)t, \quad \alpha < 0 \tag{6.72}$$

which are valid for any θ. Since ϵ can be arbitrarily small it is apparent that the trajectory approaches a logarithmic spiral as it nears the origin. Furthermore, the direction of radial motion is determined by the sign of α so a focus in the linear system corresponds to a focus in the nonlinear counterpart and the stability is the same for each.

If the linear system has vortical movement about a center then $\alpha = 0$ and Equation 6.71 reduces to

$$\frac{dR}{R} = H \, dt \tag{6.73}$$

In this case there is no intrinsic general property of the function on the right which permits a prediction of the exact nature of the trajectory. The dominant linear term is absent. Integration along a portion of the path corresponding to a displacement of 2π in θ may produce no net change in R. In this case there is a center in the nonlinear system also. However, the dominant nonlinear term may have an average sense which is either positive or negative and there will be a change in R. Therefore a center in the linear system can correspond to either a center or a focus in the related nonlinear system and the focus can be either stable or unstable. This completes the proof of the theorem of Liapunov.

The method used to discuss vorticose motion is quite general and can be used to prove the theorem for the other types of singular points. It is left as an interesting and instructive exercise to use linear transformations and conversion to polar coordinates as a means for establishing the theorem for aperiodic motion rather than the geometric proof which is given earlier.

It is appropriate now to present some examples of the forms of trajectories which can appear in nonlinear systems. The first two examples involve basically linear systems wherein several modes of operation occur. Linear analysis can be applied with only slight modifications. In the last three examples nonlinear systems which contain no linear terms are considered in order to show unusual trajectories which are not found in linear systems.

The first example is the somewhat classical treatment of Coulomb friction. This is a dynamic friction wherein the force is of constant magnitude and in the direction opposing the velocity vector. For a second order linear system which also contains an elastic restraint the differential equation is

$$\ddot{x} + p \operatorname{sgn} \dot{x} + qx = 0 \tag{6.74}$$

where p and q are constants and the symbol $\operatorname{sgn} \dot{x}$ means the algebraic sign of $\dot{x}$. In phase plane notation this implies the set of equations

$$\frac{d\dot{x}}{dx} = \frac{-qx - p}{\dot{x}} \qquad \dot{x} > 0$$

$$\frac{d\dot{x}}{dx} = \frac{-qx + p}{\dot{x}} \qquad \dot{x} < 0 \tag{6.75}$$

This suggests the transformation

$$u = \omega\left(x + \frac{p}{q}\right) \qquad \dot{x} > 0$$

$$v = \dot{x}$$

$$u = \omega\left(x - \frac{p}{q}\right) \qquad \dot{x} < 0 \tag{6.76}$$

$$v = \dot{x}$$

where in both cases $\omega = \sqrt{q}$. The situation now resembles that described by Equation 6.27 and shown in part (a) of Figure 6.8 except that there are two modes and each has its own singular point. When $\dot{x} > 0$ the trajectories in the transformed plane are circles around a center at $x = -p/q$. When $\dot{x} < 0$ the trajectories are also circles but the center is at $x = p/q$. Typical trajectories are shown in Figure 6.12. The initial conditions are selected such that the trajectories end at the same point in the interval between the singular points. Since they arrive from opposite directions no further motion is possible. All trajectories continue inward until they make first contact with this line segment and there they stop. In the x, $\dot{x}$ plane the trajectories are identical

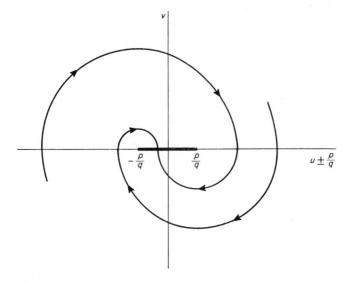

Figure 6.12 Trajectories for a system with Coulomb friction.

except for the scale factor ω along the horizontal axis. In this plane the trajectories stop where $|qx| \leqslant |p|$ or where the spring force is no greater than the frictional force.

The second example considers a system containing an inertial load only and wherein the driving torque is switched between the maximum available positive and negative values. Such systems employing either relays or switches are quite common in more complex configurations. In the simple case the differential equation is

$$I\ddot{x} = T \tag{6.77}$$

where T can assume a positive and a negative value. In the phase plane representation the equation becomes

$$\frac{d\dot{x}}{dx} = \frac{T}{I\dot{x}} \tag{6.78}$$

The variables may be separated and the solution is

$$\dot{x}^2 - \dot{x}_0{}^2 = \frac{2T}{I}(x - x_0) \tag{6.79}$$

These are parabolas symmetric to the x axis. Referring to Figure 6.13 it is assumed that the system starts at point A and it is desired to reach point B in minimum time. Two simple procedures are possible. Negative torque can be applied to reach Q and then positive torque can be applied to reach B.

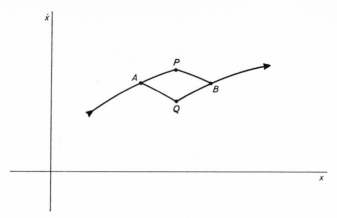

Figure 6.13   A comparison of two methods for switching torque.

Alternatively the torque values can be reversed to produce the trajectory from A to P to B. It is clear that the second procedure requires less time because everywhere the velocity is greater. Successive application of this principle shows that the quickest transition from one point to another occurs when the torque which produces maximum appropriate velocity is initially applied and maintained until the point is reached where the opposite torque produces the trajectory which passes through the desired terminal point. Usually the terminal point is a singular point representing equilibrium and when the system reaches this point the applied torque must be removed. Assuming that the singular point is translated to the origin then for the terminal portion of the trajectory Equation 6.79 gives

$$\dot{x}^2 = \frac{2T}{I}\, x \tag{6.80}$$

To represent only those trajectories which approach the origin rather than recede from it this must be changed to

$$-\dot{x}\,|\dot{x}| = \frac{2T}{I}\, x \tag{6.81}$$

and this is the locus of points in the phase plane where the relay or switch should reverse the applied torque. If this is perfectly mechanized the trajectories have the appearance shown in part (a) of Figure 6.14. If the relay or associated decision circuit has hysteresis then there are displacements of the switching line parallel to the x axis and the trajectories appear as shown in Figure 6.14 (b). There is a limit cycle. Other forms of time delay produce essentially the same result. Thermostatic temperature control is a common

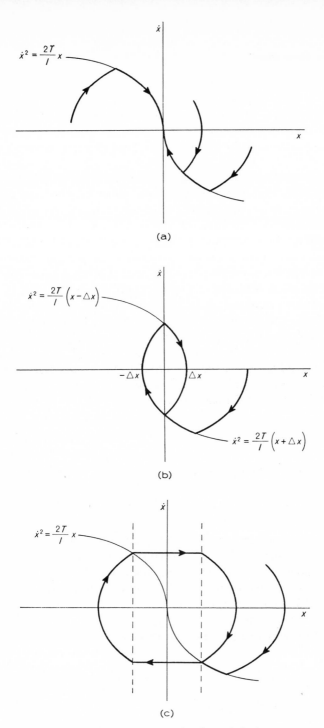

Figure 6.14 Trajectories for optimally switched system showing effects of (b) hysteresis, and (c) dead zone.

example of this type of operation wherein a terminal limit cycle is the planned mode.

If the switching circuit has a dead zone around the singular point then there is a region on both sides of the $\dot{x}$ axis wherein there is no torque. The velocity remains constant inside this region and a typical trajectory has the appearance shown in part (c) of Figure 6.14.

The above principle can be extended to systems containing damping and elastic restraint. The variety of trajectories which exist in such systems has already been demonstrated and it is apparent that optimization of switching is quite complex in such situations. It should also be mentioned that the determination of the optimum switching condition was based on an initial displacement and velocity relative to a fixed singular point. If the system is required to follow a moving equilibrium point then there is no assurance that the above criterion establishes an optimum error signal.

The last three examples show the behavior of some nonlinear systems which contain no linear terms in their characteristic. The theorem of Liapunov does not apply to such situations and the trajectories do not resemble those found in linear systems. In several cases the behavior is quite unusual and if the nonlinear characteristic could be properly mechanized the systems would make interesting toys, perhaps useful devices.

In the third example the basic equation is

$$\frac{d\dot{x}}{dx} = \frac{x^3 - 3x^2}{\dot{x}} \tag{6.82}$$

and the equation for the isoclines is

$$\dot{x} = \frac{1}{M}(x^3 - 3x^2) \tag{6.83}$$

There are singular points at $(0, 0)$ and $(3, 0)$. The trajectories are shown in Figure 6.15. It is observed that both singular points are unstable. Trajectories around the point at $x = 3$ resemble the saddle point of linear systems but the trajectories around the origin resemble nothing which has previously been discussed. This is a degenerate saddle point and in this case it is created by a particular form of nonlinear elastic restraint.

In the fourth example the basic equation is

$$\frac{d\dot{x}}{dx} = \frac{4x^2 - \dot{x}^2}{4x\dot{x}} \tag{6.84}$$

and the equation for the isoclines is

$$\dot{x} = 2[-M \pm \sqrt{M^2 + 1}]x \tag{6.85}$$

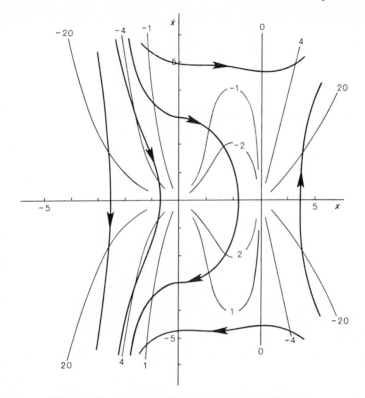

Figure 6.15   A system with two adjacent unstable points.

The only singular point is at the origin. Due to the form of Equation 6.85 all isoclines are straight lines through the origin. Some trajectories are shown in Figure 6.16. The equilibrium is unstable and the singularity is a saddle point. It is unique, however, in the behavior at $x = 0$. For motion in either direction the velocity becomes very large at this point. This is contrary to the behavior at a normal saddle point and gives an unusual appearance to the curves.

In the last example the basic equation is

$$\frac{d\dot{x}}{dx} = \frac{\dot{x}^2 - 4x^2}{4x\dot{x}} \tag{6.86}$$

and the equation for the isoclines is

$$\dot{x} = 2[M \pm \sqrt{M^2 + 1}]x \tag{6.87}$$

This is the same as the previous example except that the spring force is reversed by changing the sign of the numerator in Equation 6.84 to produce

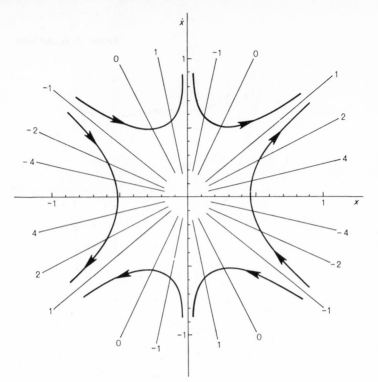

Figure 6.16 A unique saddle point.

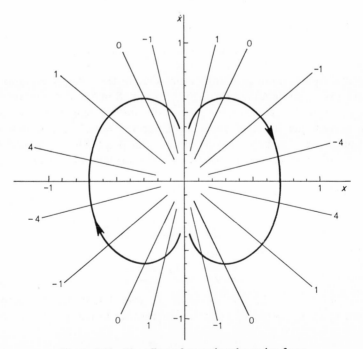

Figure 6.17 The effect of reversing the spring force.

Equation 6.86. A pair of trajectories are shown in Figure 6.17. This is most unusual behavior and it bears no resemblance to the performance of a linear system. The system is stable against displacements into the second and fourth quadrants since the motion is back to the origin. Displacements into the first and third quadrants produce motion which appears initially to be unstable but then slows to zero velocity, reverses itself, and returns to the origin. These few examples should indicate that investigations into the possible modes of behavior of nonlinear systems can be quite interesting and fructifying.

TOPOLOGY AND STABILITY

Many of the statements of behavior in the phase plane involve general relationships which are suggestive of topology although they are not usually derived by topological methods. These results are concerned with the distribution of the various types of singular points, the nature of curves which surround them, the types of trajectories which can be obtained, and relationships between trajectories. The first technique to be described has been employed by several French workers and is reported in the book by Blaquière which is listed as a reference. The method establishes the equations for a few curves which divide the phase plane into regions. The Liapunov theorem then defines the type of singular point which can be found in each region.

It is perhaps wise first to review a few aspects of linear theory. For a linear system of the most general form

$$\frac{d\dot{x}}{dt_{|}} = Ax + B\dot{x}$$

$$\frac{dx}{dt} = Cx + D\dot{x}$$

(6.88)

one assumes solutions of the form $x = x_0 e^{\lambda t}$ and $\dot{x} = \dot{x}_0 e^{\lambda t}$. A direct substitution gives

$$\lambda \dot{x} = Ax + B\dot{x}$$

$$\lambda x = Cx + D\dot{x}$$

(6.89)

and these may be written in the form

$$Ax + (B - \lambda)\dot{x} = 0$$

$$(C - \lambda)x + D\dot{x} = 0$$

(6.90)

A nontrivial solution for x and $\dot{x}$ requires that the determinant of the coefficients is zero and this gives the equation

$$\lambda^2 - (B + C)\lambda - (AD - BC) = 0$$

(6.91)

The nature of the roots and hence $x(t)$ depend on the values and relationships of the four constants in the same manner as was presented previously in the lengthy discussion on the trajectories generated by the linear Equation 6.15

$$\ddot{x} + 2p\dot{x} + qx = 0$$

In the nonlinear case the most general equations are Equations 6.57

$$\frac{d\dot{x}}{dt} = Ax + B\dot{x} + \phi(x, \dot{x}) = P(x, \dot{x})$$

$$\frac{dx}{dt} = Cx + D\dot{x} + \psi(x, \dot{x}) = Q(x, \dot{x})$$

In this case one defines the functions

$$a(x, \dot{x}) = \frac{\partial P}{\partial x}$$

$$b(x, \dot{x}) = \frac{\partial P}{\partial \dot{x}}$$

$$c(x, \dot{x}) = \frac{\partial Q}{\partial x}$$

$$d(x, \dot{x}) = \frac{\partial Q}{\partial \dot{x}}$$

(6.92)

Since ϕ and ψ contain only higher order terms the constants are found by relations such as $A = a(0, 0)$.

Since the derivatives indicate the nature of the linear portion of the system and since the linear portion essentially dictates the type of singular point this suggests plotting the three curves defined by

$$F_1(x, \dot{x}) = (b + c) = 0$$

$$F_2(x, \dot{x}) = (ad - bc) = 0$$

$$F_3(x, \dot{x}) = (b + c)^2 + 4(ad - bc) = 0$$

(6.93)

which divide the phase plane into numerous regions. The functions are zero only on the curves so they have nonzero values in the regions between curves. The various combinations of positive and negative values in a given region indicate the nature of the roots of the characteristic equation for the linear system should a singular point exist in that region. By Theorem 6.1 the same form of singular point would be found in the nonlinear system in almost all cases. The following tabulation is based on Equation 6.91 and relates the values of the F functions for the nonlinear system with the type of singularity which would be found in the corresponding linear system.

$$\left.\begin{array}{l} F_1 < 0 \\ F_3 < 0 \end{array}\right\} \to \text{Stable focus}$$

$$\left.\begin{array}{l} F_1 = 0 \\ F_3 < 0 \end{array}\right\} \to \text{Center}$$

F_2 negative
$$\left.\begin{array}{l} F_1 < 0 \\ F_3 > 0 \end{array}\right\} \to \text{Stable node}$$

$$\left.\begin{array}{l} F_1 > 0 \\ F_3 < 0 \end{array}\right\} \to \text{Unstable focus}$$

$$\left.\begin{array}{l} F_1 \geqslant 0 \\ F_3 > 0 \end{array}\right\} \to \text{Unstable node}$$

F_2 positive ------ Saddle

It should be observed that the method involves a few basic considerations. First, the functions P and Q must have linear terms to be certain that Theorem 6.1 applies. Second, the form of the derivatives must not change if a singular point is translated. Third, the results have no significance concerning a trajectory passing through a given region to a singular point in another region. A few other features may be of interest. Since it is true that

$$F_3 = F_1^2 + 4F_2 \tag{6.94}$$

whenever values of x and $\dot{x}$ make F_1 and F_2 zero simultaneously then they also make F_3 zero. This means geometrically that the F_3 curve passes through every intersection of the F_1 and F_2 curves. Also from Equation 6.94 it is true that

$$\frac{\partial F_3}{\partial \dot{x}} = 2F_1 \frac{\partial F_1}{\partial \dot{x}} + 4 \frac{\partial F_2}{\partial \dot{x}}$$

$$\frac{\partial F_3}{\partial x} = 2F_1 \frac{\partial F_1}{\partial x} + 4 \frac{\partial F_2}{\partial x}$$

(6.95)

so at a point of common intersection the fact that $F_1 = 0$ gives the relation

$$\frac{\dfrac{\partial F_3}{\partial \dot{x}}}{\dfrac{\partial F_3}{\partial x}} = \frac{\dfrac{\partial F_2}{\partial \dot{x}}}{\dfrac{\partial F_2}{\partial x}}$$

(6.96)

which means geometrically that the F_2 and F_3 curves are tangent at this point.

These and other properties are illustrated in the example shown in Figure 6.18. The basic equation for this system is

$$\frac{d\dot{x}}{dx} = \frac{-2x + 4\dot{x} - 2x^2 + \dot{x}^2}{-2x + 2\dot{x} - x^2 + \dot{x}^2}$$

(6.97)

The three requirements that $F_1 = F_2 = F_3 = 0$ ultimately reduce to the three equations

$$\dot{x} = x - 1 \qquad\qquad\qquad (F_1 = 0)$$

$$\dot{x} = \frac{1}{x} \qquad\qquad\qquad\quad (F_2 = 0) \qquad (6.98)$$

$$\dot{x} = (3x - 1) \pm 2\sqrt{2x^2 - x - 1} \quad (F_3 = 0)$$

These are plotted in Figure 6.18. The shaded area on one side of each curve indicates the side on which the function is positive. The values of the three functions in each region are then correlated with the preceding table to determine the type of singular point which might be found in each region. The functions $P = 0$ and $Q = 0$ are also plotted. Their intersections show singular points at $(0, 0)$ and $(2, 2)$.

Some isoclines and trajectories for this system are shown in Figure 6.19. The circular isocline for $M = 1.5$ is particularly interesting. It is observed that the origin is a saddle point and the point $(2, 2)$ is an unstable focus as was predicted in Figure 6.18.

If the origin is the desired equilibrium point then the saddle point is probably an undesired mode. The unstable focus is no better. Let it be assumed that a desirable system is formed by creating a singular point at a location such as the point $(3, 1)$ which would be a stable focus. This is

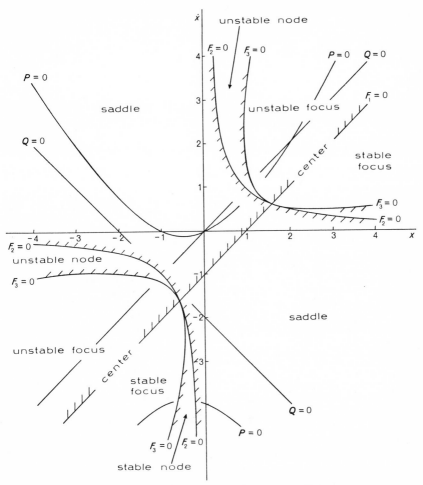

Figure 6.18 Distribution of singular points in a particular system.

accomplished by writing P and Q in the different forms

$$P' = -2(x - x_0)(x - x_1) + (\dot{x} - \dot{x}_0)(\dot{x} - \dot{x}_1) = 0$$
$$Q' = -(x - x_0)(x - x_2) + (\dot{x} - \dot{x}_0)(\dot{x} - \dot{x}_2) = 0$$

(6.99)

A comparison with the original functions shows that if the constants are chosen in accordance with the relations

$$x_0 + x_1 = -1$$
$$\dot{x}_0 + \dot{x}_1 = -4$$
$$x_0 + x_2 = -2$$
$$\dot{x}_0 + \dot{x}_2 = -2$$

(6.100)

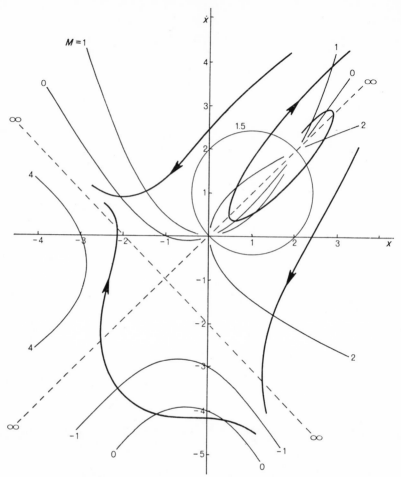

Figure 6.19 Isoclines and trajectories for the previous system.

then the F functions are unchanged but a singular point is created at the point $(x_0, \dot{x}_0)$. It should be observed that an equilibrium point which is off the x axis corresponds to a constant velocity and a fixed displacement. Such a situation appears to contradict physical reality. However, a review of the methods for finding time displacements shows that such a fixed point exists for only an infinitesimal time interval. This means that whenever a trajectory converges to a point off the x axis the phase plane presents only the transient behavior. If the trajectory actually reaches the point then $\dot{x}$ is constant and $\ddot{x} = f(x, \dot{x}) = 0$. Therefore $d\dot{x}/dx = 0$ so after the transient has decayed the dynamic system can be replaced by a static one whose trajectory is a horizontal line. This mode is frequently called a *velocity servo* and is commonly synthesized by feeding back a signal proportional to the velocity of the output rather than the position or amplitude. Such mechanization provides phase plane singular points which are off the x axis.

The next topic to be discussed is a property of a point or a curve known as its *index*. This interesting concept was introduced in a somewhat different manner by Poincaré in 1879. The index of a point whether it is singular or not is found by drawing a small closed curve enclosing the point but encircling no other singular point. For one trip around this curve one observes the rotation of the field vector defined by $d\dot{x}/dx$. Since the vector returns to its initial position the number of rotations must be an integer or zero. Using the notation of Equation 6.57 the situation is defined analytically by the relation

$$j = \frac{1}{2\pi} \oint d\left(\arctan \frac{P}{Q}\right) = \frac{1}{2\pi} \oint \frac{Q\, dP - P\, dQ}{P^2 + Q^2} \qquad (6.101)$$

Figure 6.20 shows methods for finding the indices for the various types of singular points in a linear system using the trajectories themselves as the enclosing curves. The arrows indicate regions where the field vector is either constant or nearly so. These results may be verified analytically by observing that when the linear system expressed by Equation 6.15

$$\ddot{x} + 2p\dot{x} + qx = 0$$

is converted to phase plane form then

$$P(x, \dot{x}) = -qx - 2p\dot{x}$$
$$Q(x, \dot{x}) = \dot{x} \qquad (6.102)$$

An evaluation of dP and dQ as functions of dx and $d\dot{x}$ shows that

$$\oint \frac{Q\, dP - P\, dQ}{P^2 + Q^2} = q \oint \frac{x\, d\dot{x} - \dot{x}\, dx}{P^2 + Q^2} \qquad (6.103)$$

Using a circle about the origin as the enclosing curve the signs of x, $\dot{x}$, dx, and $d\dot{x}$ are such that the integrand on the right is always positive. This leads to the conclusion that

$$\operatorname{sgn} j = \operatorname{sgn} q \qquad (6.104)$$

For a saddle point $q < 0$ and for all other singular points whether they are stable or not $q > 0$. Realizing that the field vector rotates only once in each case establishes the following.

Lemma 6.1   *The index of a saddle point is -1 and for all other singular points it is $+1$.*

It is important to observe that the sign of the index is not an indication of stability. The condition that $q < 0$ means that the elastic or spring force has

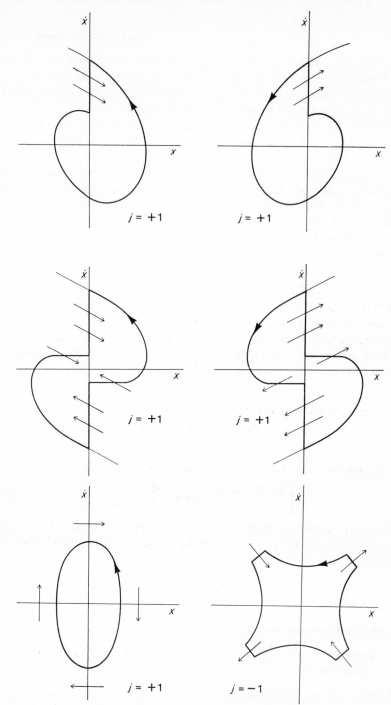

Figure 6.20 The indices in a linear system.

the same sense as the acceleration and this creates a system so unstable that no dynamic effect can stabilize it. A negative index indicates this situation. In order to extend the above linear result to the nonlinear domain a few additional properties are needed.

Lemma 6.2   *The index of a regular point is zero.*

A *regular point* is one which is not a singular point. The lemma is established by shrinking the curve around the given point. The range of the values of the isoclines can be made as small as desired. Since the index is an integer but can be made as small as desired it must be zero.

Lemma 6.3   *The index of a closed trajectory is $+1$.*

Considering the trajectory itself to be the encircling curve and recalling that the trajectory follows the field vector everywhere then this lemma follows immediately from the definition of the index.

Lemma 6.4   *A closed trajectory must surround at least one singular point.*

This is a direct consequence of Lemmas 6.2 and 6.3.

Lemma 6.5   *The index of a closed curve equals the sum of the indices of the enclosed singular points.*

This lemma serves to distinguish between the index of a point and the index of a curve. It is proved by the use of part (a) of Figure 6.21 which is reminiscent of complex variable theory. Since the curve encloses only regular points its index is zero. Decreasing the separation between the pairs of parallel lines causes their contribution to vanish. The index of the outer curve must equal the negative sum of the indices of the interior curves. Since they have a negative sense the lemma is established.

Lemma 6.6   *If the determinant of the coefficients of the linear system is not zero then the index of the singular point in the nonlinear system equals the index of its linear approximation.*

The extension theorem of Liapunov states that close to the singular point the behavior of the nonlinear system approaches that of the linear counterpart except that when the linear system has a center the nonlinear system may have a center or a focus. However, since the index is $+1$ for either a center or a focus the exception is of no consequence.

The encircling curve may be extended without changing the index. This is demonstrated using part (b) of Figure 6.21. The interior curve is sufficiently small to determine the index properly. Following the previous arguments it is observed that the outer curve may follow any path whatever provided that

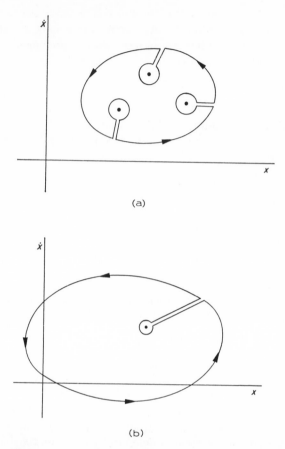

(a)

(b)

Figure 6.21 Some properties of curves surrounding singular points.

it does not encircle additional singular points. This is a very interesting property since the vector field may change considerably at large distances from the singular point yet the index remains unchanged.

If all linear terms are present and the determinant of their coefficients is zero then in general the requirements for making the numerator and denominator zero are the same and there are infinitely many singular points. When a singular point is not isolated it is called a *nonsimple singular point*. The system described by the equation

$$\frac{d\dot{x}}{dx} = \frac{x + \dot{x} + x^2}{x + \dot{x} - x^2} \tag{6.105}$$

is an interesting example of this situation. The determinant of the linear coefficients is quite obviously zero and the requirements for nulling numerator and denominator both reduce to $\dot{x} = -x$ producing an infinity of densely packed singular points along this line. The vector field requires everywhere a slope of $+1$ so the trajectories are parallel lines. The infinite set of nonsimple singular points fills the plane with regular points. Oddly, the addition of the nonlinear terms simplifies the situation. The curves $P = 0$ and $Q = 0$ are parabolas tangent at the origin to the line $\dot{x} = -x$. They intersect only at this point so this is the only singular point in the nonlinear system. When the isoclines and trajectories are drawn it is found that the behavior most closely resembles that around a saddle point except for a reversal in one part of the plane that reduces the index to zero.

The determinant also may be zero if two or more linear terms are missing. If the two terms in a row or column are zero then nonsimple singularities are again generated. Otherwise the determinant is not zero and the origin is uniquely singular. If three linear terms are zero then the determinant is necessarily zero and a coordinate axis is filled with nonsimple singularities. If all linear terms are missing then the entire plane is filled with nonsimple singularities. An interesting yet simple example of this last case is the system described by the equation

$$\frac{d\dot{x}}{dx} = \frac{-2x\dot{x}}{x^2 - \dot{x}^2} \tag{6.106}$$

If some isoclines and trajectories are drawn it is discovered that it is a saddle point with six "horns" and an index of -2.

In the previous chapter limit cycles are described as an equilibrium condition consisting of a unique mode of oscillation which can be either stable or unstable. In the phase plane this mode is represented by a closed curve which may depict either terminal vorticose motion or a limit which trajectories approach. It can be defined quite succinctly as an *isolated closed trajectory*. Therefore periodic motion of a set of vortex cycles around a center does not have a limit cycle. It is possible also to consider singular points as a degenerate form of limit cycle but that will not be done here. Further, a limit cycle is considered stable if a trajectory approaches it as $t \to \infty$ and unstable if a trajectory approaches it as $t \to -\infty$.

> **Theorem 6.2** *If the limit cycle of a trajectory contains only regular points then either the trajectory is contained in the limit cycle or approaches it from one side.*

From a topological viewpoint one of three possibilities must exist. These are shown in Figure 6.22. Either the initial conditions are such that the trajectory is entirely within the limit cycle, or it approaches it asymptotically but never

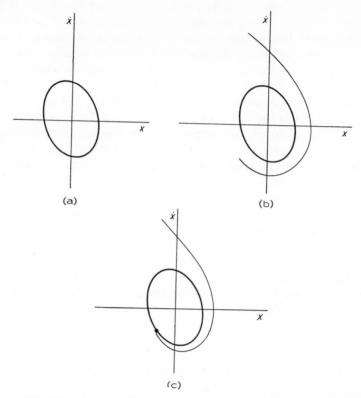

Figure 6.22   Possible relationships between a trajectory and its limit cycle.

reaches it, or at one point the trajectory enters the limit cycle. In the last case the transition point lies on the limit cycle but since at this point the paths have at least two directions the point must be singular. This violates the hypothesis so only the first two cases are permitted. The corollary follows directly from a similar argument.

Corollary 6.2.1   *If a limit cycle contains no singular points it consists of a single closed curve.*

The *half path* of a given point in the phase plane is the complete trajectory through that point as time becomes infinite. Assuming uniqueness of the vector field there are two half paths for each point, corresponding to passage to infinity in the positive and negative directions.

Theorem 6.3   *If a closed region contains a half path but no singular points inside or on its boundary then it contains also a closed path.*

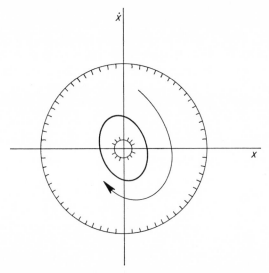

Figure 6.23   Existence of a limit cycle within a region.

Since the region is finite the half path must spiral around a region of singularity. If the given closed region contains no singularities then it must be annular in shape to exclude the singular region and the half path must spiral to a limit cycle. Since the half path is totally within the region the limit cycle must lie either inside or on the boundary. Hence it contains no singular points and by Corollary 6.2.1 it must be a closed curve. The situation is shown in Figure 6.23. An equivalent sufficient condition is stated in the following corollary.

> *Corollary 6.3.1*   *If an annular region contains no singular points and is bounded by two closed curves such that trajectories intersect the outer boundary moving inward and intersect the inner boundary moving outward then the region contains a limit cycle.*

In the previous chapter it was shown that limit cycles tend to be alternately stable and unstable. This concept can be verified and refined in the phase plane.

> *Theorem 6.4*   *For trajectories originating between two consecutive limit cycles one limit cycle is stable and the other is unstable.*

It is convenient to prove this theorem by first assuming it false. It is then possible to draw one revolution of each of two interior trajectories and to form each into a closed curve by adding a radial line segment between the endpoints. These two closed curves define a new annular region which contains half paths but no singular points. By Theorem 6.3 it must contain a closed path. Since both trajectories either approach or recede from this closed path on each side the path must be a limit cycle. This violates the fact that the original limit cycles are consecutive so the assumption is false and the theorem must be true. It should be observed that if the trajectories approach one limit cycle and recede from the other then they may spiral inside each other making the above construction impossible.

A few of the above theorems will be illustrated by an example. By means of the substitution

$$r^2 = x^2 + \dot{x}^2 \tag{6.107}$$

the general system defined by the equations

$$\frac{d\dot{x}}{dt} = \dot{x}f(x^2 + \dot{x}^2) + x$$

$$\frac{dx}{dt} = xf(x^2 + \dot{x}^2) - \dot{x} \tag{6.108}$$

reduces to the simple form

$$\frac{dr}{dt} = rf(r^2) \tag{6.109}$$

The function $f(r^2)$ can have a form such as

$$f(r^2) = \frac{g(r^2)}{\sqrt{r^2}} \tag{6.110}$$

and the differential equation reduces to

$$\frac{dr}{dt} = g(r^2) \tag{6.111}$$

The function $g(r^2)$ is completely arbitrary. Two interesting cases are shown in Figure 6.24. In part (a) the function has roots at a and b. Since dr/dt is zero at these points two circular limit cycles form part of the set of solutions of the differential equation. If $r^2 < a$ or $r^2 > b$ then dr/dt is negative. Trajectories inside the inner circle spiral to the origin and trajectories outside the outer circle spiral inward toward it. However, in the annular region between the two circles dr/dt is positive so all trajectories in this region spiral outward. As a result the origin is a stable focus, the inner circle is an unstable limit cycle, and the outer circle is a stable limit cycle. This is in consonance with Theorem 6.4. Since the boundaries of the region are limit cycles Theorem 6.3 is satisfied.

In part (b) of Figure 6.24 the function $g(r^2)$ has a root of even order at $r^2 = a$. Since $g(a) = 0$ there is a single circular limit cycle at this point.

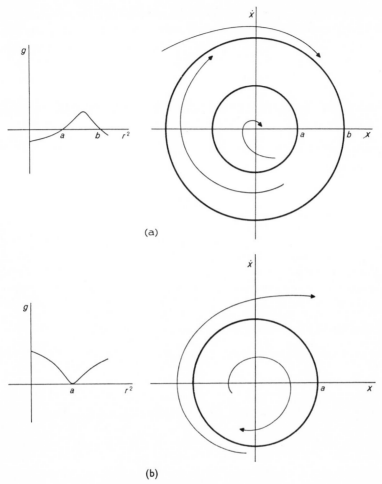

(a)

(b)

Figure 6.24   Stability of limit cycles.

However, dr/dt is positive everywhere else. This means that trajectories both inside and outside the circle spiral outward. The circle therefore is a stable limit cycle for trajectories inside it and an unstable limit cycle for trajectories outside it. A mechanical analogy would be a system having a narrow plateau of potential energy with a hill on one side and a valley on the other.

The previous theorems give some general methods for determining the existence of limit cycles in a region. Following is an analytic criterion.

Theorem 6.5   *The divergence of the vector field must change sign or be zero inside a limit cycle.*

In the last chapter a relation between line and surface integrals is established. It is known as Green's theorem and can be written as

$$\oint_C (P \, dx - Q \, d\dot{x}) = \iint_S \left(\frac{\partial P}{\partial \dot{x}} + \frac{\partial Q}{\partial x} \right) dx \, d\dot{x} \qquad (6.112)$$

If the curve C is a limit cycle of the general system

$$\frac{d\dot{x}}{dx} = \frac{P(x, \dot{x})}{Q(x, \dot{x})} \qquad (6.113)$$

then along the curve it is everywhere true that

$$P \, dx - Q \, d\dot{x} = 0 \qquad (6.114)$$

so the integrand along the curve is everywhere zero. The integral on the right can be zero only if the divergence (which is the integrand) either changes sign or is zero everywhere within the limit cycle. This is a necessary condition for the existence of a limit cycle. It is more frequently stated as a sufficient condition for the nonexistence of a limit cycle. This is done in the following corollary due to Bendixson.

Corollary 6.5.1   *If the divergence of the vector field has a fixed sign and is not zero in a region then no limit cycle can exist in the region.*

TRAJECTORIES WITHOUT ISOCLINES

Many times it is impossible to solve Equation 6.6 for either x or $\dot{x}$. A direct plot of the isoclines is then impossible. As a crude substitute it is always possible to insert arbitrary values of x and $\dot{x}$ into Equation 6.5 and obtain values of M over the phase plane. Using interpolation it is then possible to pass smooth curves through points corresponding to constant values for M. This method is applicable wherever $f(x, \dot{x})$ can be evaluated but it is tedious. Fortunately there are several methods for drawing trajectories without first plotting a field of isoclines.

In a method first proposed by Poincaré in 1885 and presented later by Szegö (whose report is listed as a reference) various families of arbitrary curves are established in the phase plane and these are then related to the trajectories by determining the locus of points where the slope of the trajectory is the same as the slope of the arbitrary curve. The curves should form a suitable grid over the phase plane but otherwise are selected for convenience and simplicity. Such a family of curves can be represented in general by

$$v(x, \dot{x}) = c \qquad (6.115)$$

where varying values of the constant c yield different members of the family. When such a family is established the gradient at any point is perpendicular to the curve passing through that point. This fact can be used to determine the desired locus. In the two-dimensional space of the phase plane the slope of the gradient must satisfy the relation

$$\frac{\dfrac{\partial v}{\partial \dot{x}}}{\dfrac{\partial v}{\partial x}} = \frac{-1}{\dfrac{d\dot{x}}{dx}} = \frac{-\dot{x}}{f(x,\dot{x})} \tag{6.116}$$

This can be simplified to

$$\frac{\partial v}{\partial x}\dot{x} + \frac{\partial v}{\partial \dot{x}}f(x,\dot{x}) = 0 \tag{6.117}$$

The substitution of the derivatives of $v(x,\dot{x})$ produces a purely algebraic expression which is the equation of the desired locus. Equation 6.117 may also be derived by setting the total time derivative of $v(x,\dot{x})$ equal to zero and substituting for $\ddot{x}$ from the original nonlinear differential equation. The table below presents some common and useful functions for $v(x,\dot{x})$ and shows the corresponding equation for the locus.

Function	Locus
x	$\dot{x} = 0$
$\dot{x}$	$f(x,\dot{x}) = 0$
$x^2 + \dot{x}^2$	$x + f(x,\dot{x}) = 0$
$x^2 - \dot{x}^2$	$x - f(x,\dot{x}) = 0$
$x^n \pm \dot{x}^m$	$nx^{n-1} \pm m\dot{x}^{m-2}f(x,\dot{x}) = 0$
$x\dot{x}$	$\dot{x}^2 + xf(x,\dot{x}) = 0$
$x^n\dot{x}^m$	$n\dot{x}^2 + mxf(x,\dot{x}) = 0$
$(x - x_0)^2 \pm (\dot{x} - \dot{x}_0)^2$	$(x - x_0)\dot{x} \pm (\dot{x} - \dot{x}_0)f(x,\dot{x}) = 0$
$Mx - \dot{x}$	$M\dot{x} - f(x,\dot{x}) = 0$

The first set of equations merely states the fact that trajectories are vertical when crossing the x axis. The second set utilizes the zero slope isocline. The last set contains the family of parallel lines with slope M and the locus which is the isocline for slope M. This combination corresponds to the usual solution by isoclines.

The example to illustrate this method will be the system described by Equation 6.82

$$\frac{d\dot{x}}{dx} = \frac{x^3 - 3x^2}{\dot{x}} \tag{6.118}$$

which was analyzed by the isocline method to give the trajectories shown in Figure 6.15. The following table lists the functions which are used in this method and shows the corresponding equation for the locus.

	Function	Locus
1.	$\dot{x} = c$	$x^2(x - 3) = 0$
2.	$x^2 + \dot{x}^2 = c$	$x(x^2 - 3x + 1) = 0$
3.	$x^2 - \dot{x}^2 = c$	$x(x^2 - 3x - 1) = 0$
4.	$x\dot{x} = c$	$\dot{x}^2 + x^4 - 3x^2 = 0$
5.	$(x + 1)^2 + \dot{x}^2 = c$	$x^3 - 3x^2 + x + 1 = 0$

In the last set of equations the value of x_0 was selected to give a root at $x = 1$ for the equation of the locus. This arbitrary translation is a useful technique for providing curves in otherwise empty regions. Furthermore it is quite easy to construct a family of circles with a common center so this technique should be employed as frequently as is feasible. All but one of the equations for the loci are functions of x only. This means that the relation is valid for all $\dot{x}$ and the graph of the locus is a vertical straight line passing through the root of the function on the x axis. Figure 6.25 shows the loci and a few

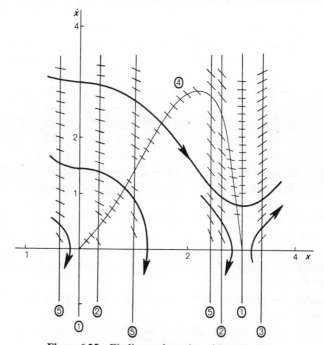

Figure 6.25 Finding trajectories without isoclines.

trajectories which were sketched from them. The short lines crossing the loci are sections from the curves determined by the corresponding functions. These short sections indicate the slope of the trajectories along the loci. The encircled numbers indicate the appropriate set of equations in the preceding table. In general the results obtained by this method and the isocline method require essentially the same amount of labor.

Figure 6.26 shows some additional methods for plotting trajectories without first drawing a set of isoclines. The first is called the *delta method* and was reported by L. S. Jacobsen in 1952. It is a very general technique and solves a broad range of problems but it is not entirely graphical. The basic idea is to define a new function

$$\delta(x, \dot{x}) = f(x, \dot{x}) + x \qquad (6.119)$$

and the original differential equation becomes

$$\frac{d\dot{x}}{dx} = \frac{\delta(x, \dot{x}) - x}{\dot{x}} \qquad (6.120)$$

The value of δ is calculated at each point P from which an increment of trajectory is to be projected. This provides the information to locate point Q. The segment of trajectory is extended from P perpendicular to the line PQ. Its slope is therefore the negative reciprocal of the slope of PQ. This is the value stipulated by Equation 6.120.

In a report suggested as a reference for this chapter Ludecke and Weber present a technique for improving the delta method. They propose the use of an average value for δ either to reduce error or to permit the use of larger increments. In general such an average value is defined by

$$\delta_{ave} = f_{ave} + x_{ave} \qquad (6.121)$$

where the mean values are calculated over the intervals for x and $\dot{x}$ which correspond to the increment of trajectory. The problem of determining a proper mean value for f is considerably reduced if it is a function only of x. Then it is convenient to use integration to find the average value and the corresponding value for δ_{ave} is given by

$$\delta_{ave} = \frac{1}{\Delta x} \int_{x}^{x+\Delta x} f(x)\, dx + \left(x + \frac{\Delta x}{2}\right) \qquad (6.122)$$

Again this must be repeatedly calculated along the trajectory. If f is a function of both x and $\dot{x}$ then δ_{ave} is a function of $\Delta \dot{x}$ but $\Delta \dot{x}$ depends on δ_{ave}. This indicates an iterative procedure. To simplify the calculation in this case it is

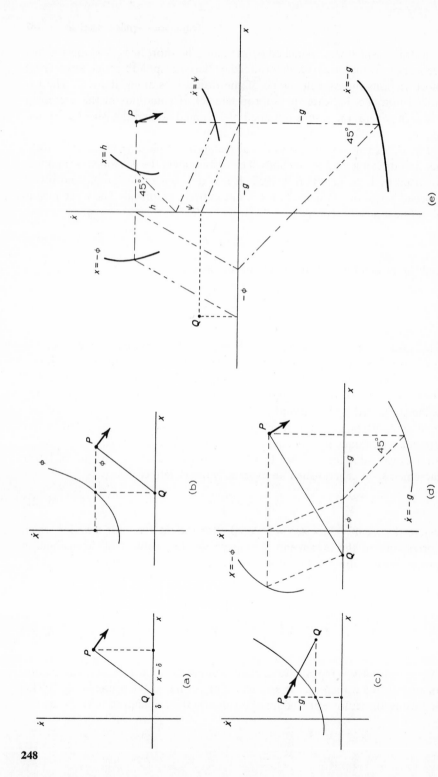

Figure 6.26 (a) Delta method. (b) Liénard method. (c) Modified Liénard method. (d) Pell method. (e) Generalized Pell method.

probably better to use linear prediction. An appropriate expression is

$$\delta_{ave} = \delta + \frac{\Delta\delta}{2}$$

$$= \delta + \frac{1}{2}\left(\frac{\partial\delta}{\partial x}\Delta x + \frac{\partial\delta}{\partial \dot{x}}\Delta\dot{x}\right)$$

$$= \delta + \frac{1}{2}\left[\left(\frac{\partial f}{\partial x} - 1\right)\Delta x + \frac{\partial f}{\partial \dot{x}}\Delta\dot{x}\right] \qquad (6.123)$$

Part (b) of Figure 6.26 illustrates a method which is entirely graphical and which provides solutions quickly and easily. The technique antedates the delta method and was first described in somewhat different form in 1928 by Liénard in a very noteworthy report which is the subject of further discussion later in this chapter. The scope of the problems which may be solved by the Liénard method is unfortunately limited to those wherein $f(x, \dot{x})$ has a special form permitting the basic differential equation to be written as

$$\frac{d\dot{x}}{dx} = \frac{\phi(\dot{x}) - x}{\dot{x}} \qquad (6.124)$$

A single curve is drawn in the phase plane. This curve is a plot of $x = \phi(\dot{x})$ and is therefore the zero slope isocline for this system. The increment of trajectory is again drawn from P perpendicular to line PQ. Reference to the figure shows that this is the slope required by Equation 6.124. Part (c) shows how the Liénard method can be modified to treat the case when $\dot{x}$ and a function of x appear in the numerator. The original differential equation then has the form

$$\frac{d\dot{x}}{dx} = \frac{g(x) - \dot{x}}{\dot{x}} \qquad (6.125)$$

In this case the curve $\dot{x} = g(x)$ is plotted. Drawing the short section of trajectory along the line joining P and Q provides the proper slope.

A significant generalization of the Liénard method was proposed in 1957 by W. H. Pell. His method treats the quite general system described by

$$\frac{d\dot{x}}{dx} = \frac{g(x) + \phi(\dot{x})}{\dot{x}} \qquad (6.126)$$

Two curves are drawn in the phase plane. These correspond to $x = -\phi(\dot{x})$ and $\dot{x} = -g(x)$. The segment of length $-g$ along the x axis is determined by the 45° projection as shown. The segment of length $-\phi$ is found by a parallel projection from the proper point on the curve $x = -\phi(\dot{x})$. The increment of trajectory is drawn from point P perpendicular to PQ.

The Pell method can be generalized as shown in part (e) of Figure 6.26. This probably represents the furthest generalization that still permits a purely geometrical solution. It treats systems of the form

$$\frac{d\dot{x}}{dx} = \frac{g(x) + \phi(\dot{x})}{h(\dot{x}) + \psi(x)} \tag{6.127}$$

and therefore provides reasonably rapid solutions to all forms of systems except those containing cross-product terms in x and $\dot{x}$. Four curves must be drawn initially. The remaining construction is merely a two-dimensional duplication of the simpler Pell method. To keep the figure as simple as possible the line PQ was not drawn but it is part of the construction since the increment of trajectory is again perpendicular to it.

LIÉNARD PLANE

In the report by Liénard to which reference has already been made a rather thorough analysis is performed on the equation

$$\ddot{x} + u(x)\dot{x} + x = 0 \tag{6.128}$$

This expression arose from the description of oscillatory physical phenomena. He introduced a transformation and various graphical and analytical procedures that were quite fruitful. He was either unaware of or indifferent to the fact that his methods are easily extended to a much larger class of systems. In deference to his original work the equation

$$\ddot{x} + u(x)\dot{x} + v(x) = 0 \tag{6.129}$$

bears his name. It is observed that this is a quite general form, requiring only that the velocity appear in a linear fashion. This equation is equivalent to the system

$$\dot{x} = y - \int_0^x u(x)\, dx = y - F(x) \tag{6.130}$$

$$\dot{y} = -v(x)$$

since forming the time derivative in the first equation and then substituting the second yields the original equation. The x, y plane is called the *Liénard plane*. Everywhere the trajectories have a slope given by

$$\frac{dy}{dx} = \frac{v(x)}{F(x) - y} \tag{6.131}$$

The form of this expression suggests that it might be possible to draw trajectories without first plotting Liénard plane isoclines. This is true. First,

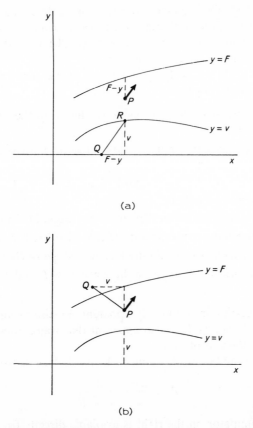

Figure 6.27 Two graphical methods for solving the
general Liénard equation.

assuming only that $u(x)$ is integrable it is possible to calculate $F(x)$ either in
closed form or by successive numerical integration. Then the curve $y = F(x)$
can be drawn. This curve is the *infinite slope isocline*. In the Liénard plane
trajectories rarely have infinite slope when crossing the x axis. Their slope
at this point is given by v/F.

It is necessary also to draw the curve $y = v(x)$ to complete the preliminary
construction. The intersections of this curve with the x axis occur at points
where $v = 0$ and this is necessary for zero slope in the trajectories. Since this
condition is independent of y the zero slope isoclines are vertical straight
lines passing through the points where $v = 0$. It may assist in the development
of solutions to draw the zero slope isoclines also.

Figure 6.27 shows two methods for finding the slope of the trajectory at
point P. In part (a) the distance from P to the curve $y = F(x)$ is used to

determine point Q. The increment of trajectory is drawn parallel to line segment QR. In part (b) the additional construction involves measuring a horizontal distance equal to v to determine point Q. The incremental trajectory is drawn perpendicular to line segment PQ.

The Liénard plane has striking similarities to the phase plane. Following are some interesting properties:

1. The intersection of a zero slope isocline with the curve $y = F(x)$ is a singular point.

2. Every closed curve must enclose area on both sides of the curve $y = F(x)$.

3. A closed curve in the Liénard plane is also a closed curve in the phase plane.

4. The origin in the Liénard plane is the origin in the phase plane.

5. A fixed point in the Liénard plane is also fixed in the phase plane.

6. Trajectories cross the infinite and zero slope isoclines alternately.

7. Trajectories move to the right when $y > F$ and to the left when $y < F$.

The above properties are intuitively apparent. Precise proofs are left as an exercise. From the above it is also apparent that trajectories have the same general form in the two planes.

Time intervals can be calculated in the Liénard plane by using the relations

$$dt = \frac{dx}{\dot{x}} = \frac{dx}{y - F} \tag{6.132}$$

where the denominator on the right is available directly from the graphical construction. The technique is shown in Figure 6.28. It is again necessary to plot a reciprocal function and perform a numerical integration.

It is also possible to calculate time intervals from the relation

$$dt = -\frac{dy}{v(x)} \tag{6.133}$$

This has the slight advantage that $1/v$ may be plotted initially. The integration, however, is to the y axis for values of y which correspond to the trajectory and these must be plotted as the construction progresses. A typical situation is shown in Figure 6.29. Since the zero slope isocline lies between points where the trajectory crosses the curve $y = F(x)$ the functions involved in this and the preceding method are infinite at quite different locations and, as usual, a combination of the two methods is likely to give best results.

By placing restrictions on the types of functions which may appear for $u(x)$ and $v(x)$ the existence and stability of limit cycles can be established. It is

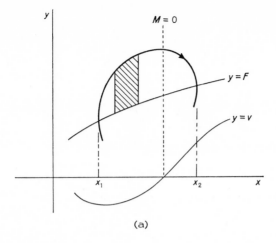

(a)

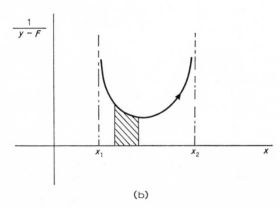

(b)

Figure 6.28 Determining the time interval in the Liénard plane.

assumed that $u(x)$ is even and $v(x)$ is odd so

$$u(x) = u(-x)$$
$$v(x) = -v(-x)$$

(6.134)

These conditions are not unduly restrictive. The functions also must be continuous and single valued. Under these assumptions if $x(t)$ is a solution then direct substitution shows that $-x(t)$ is also a solution. Since $-x$ implies $-\dot{x}$ and $-\ddot{x}$ each trajectory in the x, $\dot{x}$ plane has a complementary trajectory which is symmetric relative to the origin. In the Liénard plane

$$F(x) = \int_0^x u(x)\, dx$$

(6.135)

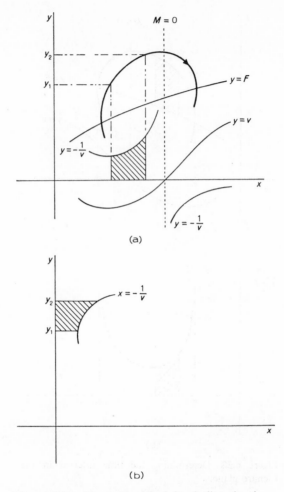

Figure 6.29 Alternate method for finding the time interval in the Liénard plane.

so $F(x)$ is odd. A substitution of $-x$ for x shows that the same symmetry property exists. The following lemma is due to this property.

Lemma 6.7   *In a Liénard system if $F(x)$ and $v(x)$ are odd, continuous, and single valued then a limit cycle exists if and only if the trajectory intersects the positive and negative y axes at equal distances from the origin.*

This is established by assuming that there exists a closed trajectory which intersects the y axis at unequal distances from the origin. Due to the symmetry property there must exist another trajectory which also intersects at unequal distances except that the intercepts are reversed. For this to happen it is necessary that the trajectories themselves intersect in at least two points. This is impossible since $F(x)$ and $v(x)$ are assumed to be single valued and the slope of every trajectory is defined uniquely at each point.

The above lemma stipulates a geometric requirement on the trajectories but it does not define the conditions under which such a trajectory can exist. The following theorem places a requirement on $F(x)$ wherein it is possible to prove that a limit cycle exists.

Theorem 6.6   In a Liénard system if $F(x)$ and $v(x)$ are odd, continuous, and single valued, and if $F(x)$ has a single positive root and increases monotonically past the null point then there exists one and only one limit cycle.

In addition it is required that $v(x)$ have a sense such that $xv > 0$. To be completely rigorous any proof of a theorem on the existence of solutions to a differential equation should invoke what is called the *Lipschitz condition*. This is a sufficient mathematical requirement for the existence and uniqueness of solutions. It essentially requires that the nonlinear functions in the equation have bounded variation for finite displacements in the independent variables. This requirement is almost automatically met in the characteristics of realizable systems so it is assumed that $F(x)$ and $v(x)$ produce a Lipschitzian situation. The proof of the theorem is based on considerations of energy. In Equation 6.129

$$\ddot{x} + u(x)\dot{x} + v(x) = 0$$

the circulating energy is given by

$$W = \int_0^x [\ddot{x} + v(x)]\, dx = \frac{\dot{x}^2}{2} + \int_0^x v(x)\, dx \qquad (6.136)$$

since the middle term gives the energy dissipated or supplied from a source. The change in circulating energy is expressed in Liénard coordinates by the relations

$$\frac{dW}{dt} = \frac{\partial W}{\partial x}\,\dot{x} + \frac{\partial W}{\partial y}\,\dot{y}$$

$$= v(x)\dot{x} + y\dot{y}$$

$$= -v(x)F(x) \qquad (6.137)$$

Direct substitutions give

$$dW = F(x)\, dy = \frac{-v(x)F(x)}{y - F(x)}$$

$$\Delta W = \int_{y_1}^{y_2} F(x)\, dy = \int_{x_1}^{x_2} \frac{-v(x)F(x)}{y - F(x)}\, dx \qquad (6.138)$$

The integration of $F(x)$ is performed along the y axis. Figure 6.30 shows how this is accomplished. The curves to the left of the y axis are plots of $x = F(x)$ using values of x taken from the trajectory. In part (a) the maximum value x_m is less than the root x_0. In this case both $F(x)$ and dy are negative at all times and hence $\Delta W > 0$. Energy is taken from the source and the amplitude increases. Due to symmetry the same is true in the other half of the cycle and a limit cycle certainly does not exist in this region.

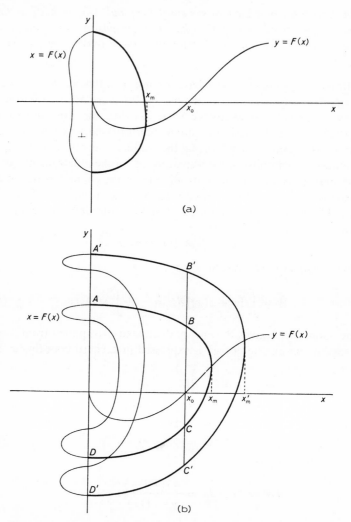

Figure 6.30 Energy considerations in a Liénard system.

Part (b) shows the more general situation where $x_m > x_0$. The trajectories must meet a few requirements. Since $v(0) = 0$ the y axis is a zero slope isocline. Since $xv > 0$ for all $x \neq 0$ the origin is the only singular point.

It is convenient to divide the energy integrals into the positive and negative components

$$\Delta W_1 = \int_{AB} dW + \int_{CD} dW > 0$$
$$\Delta W_2 = \int_{BC} dW < 0 \tag{6.139}$$

As the size of the trajectory increases, the values of x used in calculating ΔW_1 remain the same but the values of y increase. The right hand side of Equation 6.138 shows that

$$\Delta W_1' < \Delta W_1 \tag{6.140}$$

for any larger trajectory. Due to the monotonic property of $F(x)$ a reference to part (b) of Figure 6.30 shows that the area of the negative component calculated from the curve $x = F(x)$ to the y axis must increase also so

$$|\Delta W_2'| > |\Delta W_2| \tag{6.141}$$

It is not difficult to show from geometric considerations that the area of the negative component increases without bound. The change in energy ΔW is therefore a monotonically decreasing function of x_m when $x_m > x_0$. It has been shown that $\Delta W > 0$ for $x_m \leqslant x_0$ so there must be one and only one value for x_m such that $\Delta W = 0$. At this point there is neither a loss nor a gain in energy during each cycle and a limit cycle can exist.

Corollary 6.6.1   *The limit cycle in a Liénard system is stable.*

If the amplitude of oscillation should change making $\delta x_m > 0$ then this is accompanied by a change in energy $\delta W < 0$. Equation 6.136 shows that this requires a decrease in x causing the amplitude to return to the original value. The converse is true for $\delta x_m < 0$.

EXTENSIONS

In all of the preceding methods the system was of a type called *autonomous* which means that its behavior is not subject to external influence. It was also assumed that the parameters of the system were stationary in time. It is possible to construct phase plane trajectories when these restrictions are removed. The general equation for such a system is

$$\ddot{x} + g(x, \dot{x}, t) = h(t) \tag{6.142}$$

where the variation of the parameters in time is described by $g(x, \dot{x}, t)$ and where the external influence is described by the forcing function $h(t)$. This equation can be written as

$$\ddot{x} = f(x, \dot{x}, t) \tag{6.143}$$

which can be converted to the phase plane form

$$\frac{d\dot{x}}{dx} = \frac{f(x, \dot{x}, t)}{\dot{x}} \tag{6.144}$$

It is impractical to try to solve this equation using isoclines or purely geometrical methods since a new set of curves must be drawn at each point where a section of trajectory is added. Probably the most practical method for solving Equation 6.143 is a simplification of the delta method described in a preceding section. The method is illustrated in Figure 6.31(a) and is based on Equation 6.12 and Figure 6.5 where it was shown that the projection to the x axis of the normal to the trajectory equals the acceleration. Given initial

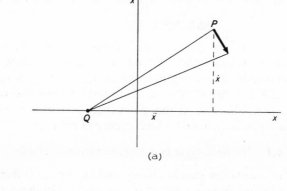

(a)

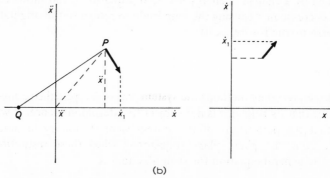

(b)

Figure 6.31 (a) Time dependent system. (b) Third-order system.

values of x, $\dot{x}$, and t the corresponding value of $\ddot{x}$ may be calculated from Equation 6.143. This is used to locate point Q. The figure shows a negative value for $\ddot{x}$. The increment of trajectory is drawn perpendicular to the line PQ. The time interval equals the angle subtended from Q by the increment of trajectory. In this case the calculation of time intervals is no longer optional. A running computation of time is necessary to compute $\ddot{x}$ at each point to determine the new location of the point Q.

All of the preceding methods treat only second-order systems. The techniques may be extended to systems of higher order with a corresponding increase in complexity. The phase plane is replaced with a *phase space* where the number of dimensions equals the original order of the system. The general method is illustrated with a third-order system but it is quite apparent that the procedure can be extended indefinitely.

The general equation for a third-order system is

$$\dddot{x} = f(x, \dot{x}, \ddot{x}, t) \tag{6.145}$$

This is in essentially the same form as Equation 6.143 and it can be expected that a similar method can be used. The procedure is shown in part (b) of Figure 6.31. It is based on the fundamental relation

$$\frac{d\ddot{x}}{d\dot{x}} = \frac{\dddot{x}}{\ddot{x}} \tag{6.146}$$

The value of $\dddot{x}$ is calculated from Equation 6.145 and used to locate point Q in the $\dot{x}$, $\ddot{x}$ plane. The figure shows a negative value for $\dddot{x}$. Drawing the increment of trajectory perpendicular to line PQ provides the slope required by Equation 6.146. This is only part of the solution. To find x it is necessary to plot a trajectory in the x, $\dot{x}$ plane also. This is based on the relation

$$\frac{d\dot{x}}{dx} = \frac{\ddot{x}}{\dot{x}} \tag{6.147}$$

The line in the $\dot{x}$, $\ddot{x}$ plane from the origin to point P has exactly this slope. The initial point in the x, $\dot{x}$ plane is either given or it is the result of previous construction so the orientation of the trajectory is determined. The terminal point $\dot{x}_1$, is found directly from the $\dot{x}$, $\ddot{x}$ plane.

It is interesting and important to note that all computation is performed only at the plane of highest order. This is true in all cases. The completion of the other trajectories consists only of measuring slopes and terminal points. Therefore the increase in complexity with order is not as severe as might be expected.

REFERENCES

1. S. Lefschetz, *Differential Equations: Geometric Theory*, Interscience, New York.
2. R. A. Struble, *Nonlinear Differential Equations*, McGraw-Hill, New York, 1962.
3. A. Blaquière, *Nonlinear System Analysis*, Academic Press, New York, 1966.
4. G. P. Szegö, A New Procedure for Plotting Phase Plane Trajectories, *Transactions of the AIEE*, Volume 81, Number 61, 1962, pp. 120–125.
5. L. S. Jacobsen, On a General Method of Solving Second Order Ordinary Differential Equations by Phase Plane Displacements, *Journal of Applied Mechanics*, Volume 19, 1952, pp. 543–553.
6. C. A. Ludecke and R. R. Weber, An Optimization of the Phase Plane Delta Method for the Solution of Nonlinear Differential Equations, *Quarterly of Applied Mathematics*, Volume 20, 1962, pp. 67–77.
7. A. Liénard, Etude des Oscillations Entretenues, *Revue Generale de L'Electricite*, Volume 23, 1928, pp. 901–912, 946–954.
8. R. A. Whitbeck, Phase Plane Analysis, *Information and Control*, Volume 4, 1961, pp. 30–47.
9. W. H. Pell, Graphical Solution of Single-degree-of-freedom Vibration Problem with Arbitrary Damping and Restoring Force, *Journal of Applied Mechanics*, Volume 24, 1957, pp. 311–312.

7

THE LIAPUNOV METHOD

In 1892 M. A. Liapunov published a lengthy discussion of the solution of differential equations. His report contained, among many other things, what is regarded today as the most powerful method for predicting the stability of systems described by differential or difference equations. The subsequent development of his method is due almost entirely to other Russian mathematicians but much has been translated and today there is considerable literature on Liapunov theory in the English language. Several books are listed as references for this chapter.

The method has great generality and uses some rather sophisticated and esoteric mathematical concepts. Therefore this short chapter can presume only to present the essential elements in a reasonably rigorous fashion and to apply the method to some basic situations. Before the fundamental theorems can be established, however, it is necessary to introduce some preliminary considerations.

First, the definition of phase space must be expanded. In the preceding chapter the coordinates of the phase space are the successive time derivatives of a single variable. For the present purposes it is assumed that the state of the system is described by n variables $x_1, x_2, \ldots, x_n$. These can be generalized coordinates and their velocities and accelerations. For simplicity of notation the ensemble of coordinates is designated by the single symbol x. Each of the variables is a function of time and a trajectory which describes a motion of the system is a curve in a space of n dimensions.

It is assumed that the system is described by the differential equation

$$\dot{x} = X(x, t) \tag{7.1}$$

which represents a set of equations of the form

$$\dot{x}_i = X_i(x_1, x_2, \ldots, x_n, t) \tag{7.2}$$

This is recognized as the *normal form* which is generated by the Bryant-Stern equations of Chapter 2. In most systems there is at least one set of values

261

for x and t which makes $\dot{x} = 0$. Since all time derivatives are zero this is a point of *equilibrium* in the phase space. A simple linear translation can place any equilibrium point at the origin and it is assumed that this has been done.

To simplify the statements of numerous requirements it is convenient to define some special types of functions. If the function $W(x)$ and its first partial derivatives are continuous in a region about the origin and if $W(0) = 0$ but $W(x) > 0$ elsewhere in the region then $W(x)$ is said to be *positive definite* in the region. An example is shown in Figure 7.1(a). The illustration necessarily shows a function of only one variable. If $W(x) \geqslant 0$ at points in the region

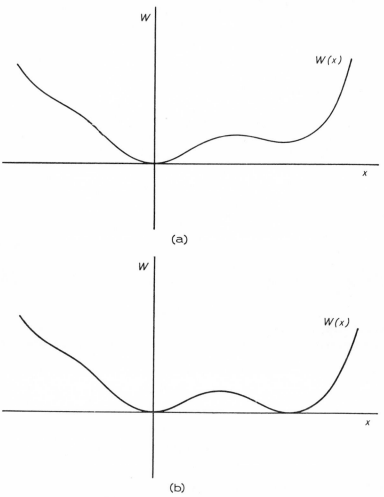

(a)

(b)

Figure 7.1   A positive definite and a positive semidefinite function.

other than the origin then the function is said to be *positive semidefinite*. This case is shown in Figure 7.1(b). There are analogous definitions for the negative sense.

If the system is not autonomous then the time t appears explicitly. In this case it is necessary that a function maintain positive definiteness uniformly in t. This is true under the following definition. A function $U(x, t)$ is *positive definite* if $U(0, t) = 0$ for all t and if there exists a stationary positive definite function $W(x)$ such that $U(x, t) \geqslant W(x)$ for all t and for all x in the region. In this case $U(x, t)$ is said to *dominate* $W(x)$ in the region. To illustrate time-dependent positive definiteness the function

$$U(x, t) = (\cos t - 1)(x_1^2 + x_2^2) \tag{7.3}$$

is not positive definite because when t equals zero or some even multiple of π no positive definite $W(x)$ can be found. If $\cos t$ is replaced with t then $U(x, t)$ is positive definite, for example, for $t \geqslant t_0 = 2$.

A function $U(x, t)$ is *positive semidefinite* if $U(0, t) = 0$ for all t and $U(x, t) \geqslant 0$ for all t and for all x in the region. This is essentially the same as the requirement for $W(x)$ and the function given by Equation 7.3 is positive semidefinite for all t.

A positive definite function whose total time derivative is negative semi-definite is called a *Liapunov function*. These are designated $V(x, t)$. The definition implies that $V(0, t) = \dot{V}(0, t) = 0$ and also that $V(x, t) > 0$ and $\dot{V}(x, t) \leqslant 0$ if $x \neq 0$. It is shown later that the essential technique in the Liapunov method consists of (a) constructing a suitable positive definite function of the variables and (b) calculating its total time derivative. It is not difficult to show that the lowest degree terms in x of $V(x, t)$ must be even if the function is to be positive definite. Quadratic forms are frequently used. A simple function of the form

$$V(x, t) = \sum_{i=1}^{n} a_i x_i^2 \tag{7.4}$$

is certainly suitable if all of the a_i are positive. The coefficients may be explicit functions of the time but only certain functions are acceptable. They must assure that $V(x, t)$ is positive definite in accordance with the previous definition and they must prevent the time derivative from becoming positive for any $t \geqslant t_0$ and for any x in the region.

Calculation of the time derivative can be indicated by the equation

$$\dot{V}(x, t) = \sum_{i=1}^{n} \dot{x}_i \frac{\partial V}{\partial x_i} + \frac{\partial V}{\partial t} \tag{7.5}$$

If the system is autonomous the time t does not appear explicitly and the time derivative can be written more concisely in vector notation as

$$\dot{V}(x) = X \cdot \text{grad } V \tag{7.6}$$

The above are the most basic considerations about Liapunov functions. At the present time there is an extensive body of literature concerning the construction of suitable Liapunov functions in various situations. The listed references give points of departure for such investigations.

STABILITY CRITERIA

In the fourth chapter some systems are said to be potentially unstable because there is enough power to overcome losses. This is a primitive and inaccurate criterion. In the fifth chapter the simple criteria for linear systems are extended to include nonlinear systems under some limiting assumptions. In the preceding chapter trajectories are considered stable if they do not leave the region of interest or desired activity. Thus, up to this point the concept of stability is based on mathematical expediency rather than a precise definition. It is the purpose of this section to discuss various possibilities and to formulate several precise definitions.

A *parameter* is a quantity which influences the type of behavior of the system but which is not in itself the subject of interest. Its value can usually be controlled independently of the motion. The range of permissible values of the parameters defines a *parameter space*. A particular point in the parameter space establishes a unique motion which is called the *unperturbed motion*. A change in parameters creates a *perturbed motion*. Although it is usually possible to change components or properties of a system the mathematical theory normally contemplates changes in the initial conditions x_0 and t_0.

If the behavior is described by a set of variables which are bounded at all times then the system is said to have *Lagrange stability*. The concept is simple and convenient. The criterion is probably adequate for linear systems but it is rather unsuitable for nonlinear systems. An atomic reactor which runs out of control, explodes, and showers debris over the surface of the earth is stable by the Lagrange criterion. The inadequacy is evident.

Liapunov defined two different types of stability and much of the present theory is based on his original definitions. These are stated first for the behavior of a trajectory relative to its equilibrium point.

The origin is said to be *stable* if for every $\epsilon > 0$ there exists a $\delta > 0$ such that if

$$|x_0| < \delta \tag{7.7}$$

then

$$|x(t)| < \epsilon \tag{7.8}$$

for all $t \geqslant t_0$. The symbol $|x|$ means the *norm* of x. This function has a standard mathematical definition which permits several interpretations but a suitable representation is the ordinary absolute value of the vector x. There is no requirement that ϵ be infinitesimal. This criterion merely states that if a

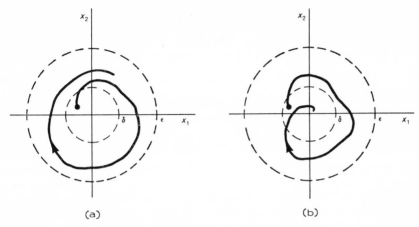

Figure 7.2 Stability and asymptotic stability.

trajectory starts at t_0 within a certain distance from the origin then it will always remain within a specified distance from the origin. A continuous oscillation is not excluded. In general δ is a function of both ϵ and t_0. The stability is said to be *uniform* if δ is independent of t_0. Figure 7.2(a) illustrates a stable motion in two dimensions.

A simple system whose behavior is described by the relation

$$x(t) = a \cos \omega t \tag{7.9}$$

is stable for $t_0 = 0$. By selecting the parameter $|a| < \delta = \epsilon$ the requirements of the definition are met. However, if the same motion is considered at $\omega t_0 = \pi/2$ then the proper value for δ cannot be determined and the motion is considered to be unstable. This consequence of the definition may be a bit difficult to accept from physical considerations. An example of a bounded, unstable motion is

$$x(t) = \sin \omega t \tag{7.10}$$

since ϵ cannot be selected less than one.

The origin is said to be *asymptotically stable* if (a) it is stable and (b) whenever

$$|x_0| < \delta \tag{7.11}$$

then

$$\lim_{t \to \infty} x(t) = 0 \tag{7.12}$$

The second part of the definition can be replaced with the equivalent requirement that given $\xi > 0$ there is a τ such that $|x(t)| < \xi$ if $|x_0| < \delta$ and $t > t_0 + \tau$. In this case the asymptotic stability is uniform (a) if the stability is uniform and (b) if τ is independent of x_0 and t_0. Asymptotic stability is illustrated in Figure 7.2(b).

Since the trajectory of an asymptotically stable system approaches the origin it is somewhat puzzling to understand why it is necessary also to require the system to be stable. This is best illustrated by an example. A system whose behavior is described by the equation

$$x(t) = ae^{-t} \cos \omega t \qquad (7.13)$$

is asymptotically stable at $t_0 = 0$. This is not true for $\omega t_0 = \pi/2$. Again $x_0 = 0$ regardless of the choice for a and the previous difficulties are encountered. However, for any finite a the trajectory will remain within a distance of ξ from the origin after a sufficiently long time interval. Thus after an excursion which may exceed the bounds for a stable system the trajectory ultimately behaves properly. This special situation is known as *quasiasymptotic stability*. As shown above it can be entirely disjoint from stability. In the language of logic asymptotic stability is the intersection of stability and quasiasymptotic stability.

The above considerations apply to motion of a single trajectory relative to an equilibrium point at the origin. They can be applied also to the relative motion of trajectories. A trajectory $x(t)$ is said to be *stable* if for every $\epsilon > 0$ there exists a $\delta > 0$ such that if, for a neighboring trajectory $x'(t)$,

then
$$|x_0 - x_0'| < \delta \qquad (7.14)$$

$$|x(t) - x'(t)| < \epsilon \qquad (7.15)$$

for all $t \geqslant t_0$. Under this definition $x(t) = at$ is unstable but $x(t) = a + t$ is stable. A trajectory is *asymptotically stable* if (a) it is stable and (b) whenever

then
$$|x_0 - x_0'| < \delta \qquad (7.16)$$

$$\lim_{t \to \infty} [x(t) - x'(t)] = 0 \qquad (7.17)$$

The concept of uniformity is applied similarly and there is also the special case of quasiasymptotic stability.

The motion $x(t) = a \cos \omega t$ is stable at $t_0 = 0$ for variations in the parameter a. The situation is quite different for ω. If this parameter is changed any finite amount then there are times when the distance between the unperturbed and perturbed motions equals $2a$. The only way to meet the requirement for stability is to make a vanish. But the elimination of motion is not the achievement of stability.

Such a consideration led Poincaré to introduce a definition more appropriate for cyclic motions. A trajectory is said to be *orbitally stable* if for every $\epsilon > 0$ and $t > t_0$ there is a $\delta > 0$ and $t' > t_0$ such that if

$$|x_0 - x_0'| < \delta \qquad (7.18)$$

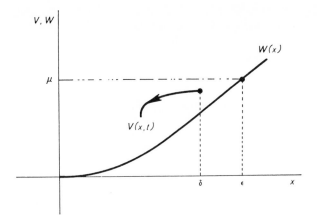

Figure 7.3 Liapunov stability theorem.

then

$$|x(t) - x'(t')| < \epsilon \tag{7.19}$$

In this case the distance between points on the two trajectories is calculated for different values for t. This definition has an interesting portrayal. One envisions a tube with a radius of ϵ surrounding $x(t)$ in a space of n dimensions. The definition states that if $x'(t)$ starts at t_0 within a distance of δ from $x(t)$ then it will remain forever within the tube.

STABILITY THEOREMS

This section introduces the original stability theorems of Liapunov. The very inspired central idea in his method is to predict the stability of the solutions of a set of differential equations by constructing an auxiliary function of the variables and by placing requirements on its behavior which permit only certain types of motion. Thus it is not necessary to solve the equations. It is not surprising that the auxiliary function is a Liapunov function $V(x)$ or $V(x, t)$. Throughout the following discussion care should be taken to distinguish between the trajectory $x(t)$ and the auxiliary function.

> **Theorem 7.1** *If a system has a Liapunov function then the origin is stable.*

Figure 7.3 illustrates the proof of the theorem for a single variable but the following proof is true for any n. By definition $V(x, t)$ dominates a positive definite function $W(x)$. It is always possible to select μ such that $W(x) \geqslant \mu$ for $|x| = \epsilon$ where ϵ confines x to an appropriate region. Since $V(x, t)$ is continuous in x and $V(0, t) = 0$ there is a $\delta < \epsilon$ such that $V(x_0, t_0) < \mu$ if $|x_0| < \delta$. For values of x and t along a trajectory

$$V(x, t) = V(x_0, t_0) + \int_{t_0}^{t} \dot{V} \, dt \qquad (7.20)$$

and since always $\dot{V} \leqslant 0$ it is true that

$$V(x, t) \leqslant V(x_0, t_0) < \mu \qquad (7.21)$$

Hence $|x|$ can never become as large as ϵ because at that point $V(x, t) \geqslant \mu$.

The proof of the theorem depends on a conflict between the positive definiteness of $V(x, t)$ and its nonpositive time derivative. Therefore both properties are necessary to provide the sufficient conditions for stability. It should also be observed that the stability determined under these conditions is precisely the type defined previously and illustrated in part (a) of Figure 7.2.

In the autonomous case the time t does not appear explicitly. There is no need for $W(x)$ and μ is selected such that $V(x) \geqslant \mu$ for $|x| = \epsilon$. The remainder of the proof is the same.

In the above situation the algebraic signs of V, $\dot{V}$, and W can be reversed and the origin remains stable. Such consistent sign reversals are possible in all of the theorems of this section but there is usually no advantage in such a procedure.

Theorem 7.2 *If a system has a Liapunov function which is dominated by another positive definite function and whose total time derivative is negative definite then the origin is asymptotically stable.*

First it should be observed that the requirements of Theorem 7.1 are satisfied and therefore the system is stable. As t becomes infinite let the limit of x be called x_f and let the limit of V be called V_f. The latter quantity can be positive or zero. First it is assumed that V_f is positive. If $V_f > 0$ then $W_1(x_f) > 0$ and $|x_f| > 0$. Let the minimum value of $x(t)$ for the entire trajectory be x_m. As shown in Figure 7.4 the presence of the dominating function $W_1(x)$ assures that $x_m > 0$. Therefore at all times

$$\dot{V}(x, t) \leqslant W_2(x) = \mu < 0 \qquad (7.22)$$

Equation 7.20 supplies the relation

$$V(x, t) \leqslant V(x_0, t_0) + \mu(t - t_0) \qquad (7.23)$$

Since μ is negative in this expression, when t exceeds a certain value then $V(x, t)$ becomes negative. But this is impossible since always $V(x, t) \geqslant 0$. Therefore the original assumption was incorrect and $V_f = 0$. Hence $x_f = 0$ and the origin is asymptotically stable.

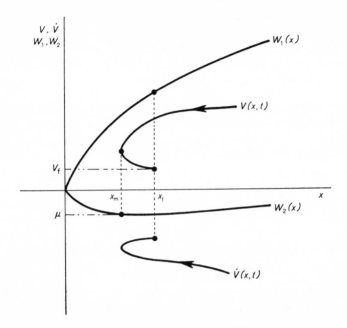

Figure 7.4   Liapunov asymptotic stability theorem.

The presence of the dominating functions in the above theorem forces $V(x, t)$ to have the proper behavior uniformly in t. An autonomous system is always uniform in t and the following simpler theorem applies.

Theorem 7.3   *If an autonomous system has a Liapunov function whose total time derivative is negative definite then the origin is asymptotically stable.*

The proof is very similar to the case where t appears explicitly except that W, W_1, and W_2 are not needed and μ can be taken directly as the minimum value of $\dot{V}$. Asymptotic stability follows immediately. Details of the proof are left as an exercise.

The above basic theorems state essentially that if an auxiliary function with stipulated properties can be found then the system is known to have some form of stability. Thus the existence of a Liapunov function is a sufficient condition. It can be proved that if the system possesses some form of stability then an appropriate Liapunov function exists and hence the condition is also necessary. Such proofs essentially give methods for constructing Liapunov functions in systems with prescribed properties. The proofs are rather complex and since the necessity is of little practical interest they are not given here.

Although a stable system must display stability for displacements in all directions a system is considered to be unstable if it is unstable for particular displacements only. This fact was recognized by Chetaev to whom the following theorem is due. The original theorems of Liapunov are weaker and are given as corollaries.

Theorem 7.4 *If a system has a function $V(x, t)$ such that $V(0, t) = 0$ and in a region bounded by $|x| = \epsilon$ and $V(x, t) = 0$ there is an upper bound for V, $\dot{V}$ is positive definite, and there are points arbitrarily near the origin where $V > 0$ for all $t > t_0$ then the origin is unstable.*

The situation for two variables is shown in Figure 7.5. Given ϵ let V_m be the maximum value of $V(x, t)$ within the region and on the boundary. Assume that $\dot{V}(x, t)$ dominates $W_2(x)$. Given an initial point (x_0, t_0) where $V > 0$ in the region let μ be the minimum value of $W_2(x)$ in the region and in the interval $|x_0| \leqslant |x| \leqslant \epsilon$. Equation 7.20 then gives the relation

$$V(x, t) \geqslant V(x_0, t_0) + \mu(t - t_0) \tag{7.24}$$

Since μ is positive there is a time t for which $V = V_m$ and the trajectory reaches or passes the boundary of the region. Since $V(x_0, t_0) > 0$ and $\dot{V} > 0$ the trajectory cannot reach the equilibrium point at the origin nor the

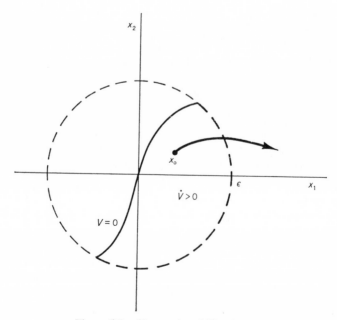

Figure 7.5   Chetaev instability theorem.

boundary $V = 0$. The motion can reach only the boundary where $|x| = \epsilon$ and the system is unstable. The first of the following two corollaries is contained in the theorem.

Corollary 7.4.1 *If a system has a function $V(x, t)$ which is positive in regions arbitrarily close to the origin, which is bounded, and which has a positive definite total time derivative for $|x| \leqslant \epsilon$ then the origin is unstable.*

Corollary 7.4.2 *If a system has a function $V(x, t)$ which is positive in regions arbitrarily close to the origin, which is bounded, and which has a total time derivative $\dot{V} = \lambda V + V'$ where $\lambda > 0$ and $V' \geqslant 0$ then the origin is unstable.*

In this corollary V' is an independent function. From the construction of $\dot{V}$ it is true that

$$\frac{d}{dt}(e^{-\lambda t}V) = e^{-\lambda t}\dot{V} - \lambda e^{-\lambda t}V$$

$$= e^{-\lambda t}V' \geqslant 0 \tag{7.25}$$

Hence along any trajectory the function V increases at a rate given by

$$V(x, t) \geqslant e^{\lambda t}V(x_0, t_0) \tag{7.26}$$

so it must eventually exceed the bound for V in any region $|x| \leqslant \epsilon$. This is possible only if it leaves the region so the system is unstable.

Before proceeding with further theoretical discussion it is perhaps wise to give some applications of the above theorems. First is a basic theorem in mechanics which was formulated by Lagrange and later proved by Dirichlet. It is easily established by the Liapunov method.

Theorem 7.5 *In a conservative system an equilibrium point where the potential energy is a minimum is stable.*

It is shown in the second chapter that in a conservative system there exists a Hamiltonian function $H(x, \dot{x})$ which is the sum of the kinetic energy and the potential energy. It is assumed that the equilibrium point is translated to the origin and that the Hamiltonian has the form

$$H(x, \dot{x}) = T(\dot{x}) + U(x) \tag{7.27}$$

where $T(\dot{x})$ is the kinetic energy and $U(x)$ is the potential energy. Reference to Equations 2.5 and 2.34 shows that the kinetic energy is positive definite with respect to $\dot{x}$. Since the potential energy is assumed to have a minimum at $x = 0$ and it is defined only to within an arbitrary constant then this can be selected to make the function positive definite. In a conservative system

$\dot{H} = 0$ everywhere so $H(x, \dot{x})$ is a suitable Liapunov function. Its time derivative is negative semidefinite so the origin is stable.

In the sixth chapter the discussion following Equation 6.50 shows how the equation of a second-order system can be written as

$$\dot{x}_1 = \lambda x_1 + f_1(x_1, x_2)$$
$$\dot{x}_2 = \mu x_2 + f_2(x_1, x_2)$$

$$(7.28)$$

where λ and μ are the roots of the characteristic equation for the linear part of the system and where f_1 and f_2 contain only terms of higher order. If λ and μ are real and have the same sign then a suitable Liapunov function is

$$V = x_1{}^2 + x_2{}^2 \qquad (7.29)$$

Its time derivative is

$$\dot{V} = 2x_1\dot{x}_1 + 2x_2\dot{x}_2$$
$$= 2(\lambda x_1{}^2 + \mu x_2{}^2 + x_1 f_1 + x_2 f_2)$$

$$(7.30)$$

If λ and μ are negative then $\dot{V}$ is negative definite in a region close to the origin and the system is asymptotically stable. The origin is a stable node. If λ and μ are positive then $\dot{V}$ is positive definite in the region and the origin is unstable.

If λ is positive and μ is negative then a suitable Liapunov function is

$$V = x_1{}^2 - x_2{}^2 \qquad (7.31)$$

This is positive in the region where $x_1{}^2 > x_2{}^2$. Its time derivative is

$$\dot{V} = 2(\lambda x_1{}^2 - \mu x_2{}^2 + x_1 f_1 - x_2 f_2) \qquad (7.32)$$

This is positive definite close to the origin in the same region where $V > 0$. The requirements of Theorem 7.4 are satisfied and the origin is unstable. It is recognized as a saddle point.

If λ and μ are the complex conjugates $a \pm jb$ then a suitable Liapunov function is

$$V = x_1 x_2 \qquad (7.33)$$

Since x_1 and x_2 are also complex conjugates V is real and positive definite. Its time derivative is

$$\dot{V} = \dot{x}_1 x_2 + x_1 \dot{x}_2$$
$$= 2aV + x_2 f_1 + x_1 f_2$$

$$(7.34)$$

If $a < 0$ then $\dot{V}$ is negative definite in a region close to the origin and the system is asymptotically stable. The origin is a stable focus. If $a > 0$ then $\dot{V}$ is positive definite in the same region and the origin is unstable. If $a = 0$ then the sign of $\dot{V}$ is determined by the terms of higher order. These results are consistent with Theorem 6.1.

In the sixth chapter there is a detailed discussion of a very general nonlinear equation known as the Liénard equation. It has the form

$$\ddot{x} + u(x)\dot{x} + v(x) = 0 \qquad (7.35)$$

It is perhaps interesting and instructive to make a comparative analysis using the Liapunov method. First it must be assumed that $u(x)$ and $v(x)$ are sufficiently well behaved to provide unique solutions to the equation. Next, under the definitions

$$x_1 = x$$

$$x_2 = -\int_0^t v(x)\, dt \qquad (7.36)$$

$$F(x_1) = \int_0^{x_1} u(x)\, dx$$

the equivalent system is

$$\dot{x}_1 = x_2 - F(x_1)$$
$$\dot{x}_2 = -v(x_1) \qquad (7.37)$$

which can be verified by differentiating the first equation with respect to time and making appropriate substitutions. If $x_1 = x_2 = 0$ and $v(0) = 0$ then $\dot{x}_1 = \dot{x}_2 = 0$ and there is an equilibrium point at the origin. As a Liapunov function it is convenient to use the energy circulating between the inertial and elastic elements when $u = 0$. This is the W of Equation 6.136. Defining the function

$$G(x_1) = \int_0^{x_1} v(x)\, dx \qquad (7.38)$$

then the Liapunov function can be written as

$$V = \int_0^{x_1} (\ddot{x} + v)\, dx$$

$$= \frac{x_2^2}{2} + G(x_1) \qquad (7.39)$$

Since $G(0) = 0$ regardless of the form of $v(x)$ then V is positive definite in a region if $x_1 v(x_1) > 0$ for all $x_1 \neq 0$ in the region. This is only a sufficient condition but it is a restriction on $v(x)$ more mild than requiring it to be an odd function which is done in the last chapter.

The time derivative of V is given by

$$\dot{V} = v\dot{x}_1 + x_2\dot{x}_2 = -vF \qquad (7.40)$$

It is observed that $\dot{V}(0) = 0$ because $v(0) = 0$ so $\dot{V}$ is negative definite if $v(x_1)F(x_1) > 0$ for all $x_1 \neq 0$ in the region. These sufficient requirements on

V and $\dot{V}$ can be expressed quite succinctly by the relation

$$\operatorname{sgn} F = \operatorname{sgn} v = \operatorname{sgn} x_1 \tag{7.41}$$

For this to happen it is sufficient but certainly not necessary that $u > 0$ for all x_1 in the region. This provides the rather obvious conclusion that a second order system is asymptotically stable if the elastic effect provides a restoring force and if the damping is always positive. In the more general case the requirement that $\operatorname{sgn} F = \operatorname{sgn} x_1$ permits $u(x)$ to become negative for certain values of x provided only that $F(x_1)$ behaves properly. If the damping term $u(x)$ is negative to the extent that $F = 0$ for some $x \neq 0$ and $F > 0$ elsewhere then $\dot{V}$ is only negative semidefinite and the system is only stable. An oscillation may exist. If the damping is negative near the origin so that F has the appearance shown in Figure 6.30 then $\dot{V}$ is positive definite near the origin and the origin is unstable. It is shown in Chapter 6 that in this case the motion is away from the origin toward a stable limit cycle.

The Liénard equation can also be used to demonstrate the determination of the stability of solutions when the time t appears explicitly. Consider the equation

$$\ddot{x} + u(x, \dot{x}, t)\dot{x} + v(x) = 0 \tag{7.42}$$

where the damping is dependent also on velocity and time. In this case it is convenient to construct the equivalent system

$$\begin{aligned} \dot{x}_1 &= x_2 \\ \dot{x}_2 &= -u(x_1, x_2, t)x_2 - v(x_1) \end{aligned} \tag{7.43}$$

The origin is an equilibrium point if $v(0) = 0$. The Liapunov function is again defined by Equation 7.39

$$V = \frac{x_2{}^2}{2} + G(x_1)$$

where $G(x_1)$ is calculated by Equation 7.38. Since $G(0) = 0$ the function V is again positive definite if $x_1 v(x_1) > 0$ for all $x_1 \neq 0$ in the region. The time derivative of V is given by

$$\dot{V} = v\dot{x}_1 + x_2\dot{x}_2 = -ux_2{}^2 \tag{7.44}$$

It is observed that $\dot{V}(0) = 0$. The negative definiteness of $\dot{V}$ therefore depends on the properties of $u(x_1, x_2, t)$. According to the definition of time-dependent positive definite functions it is sufficient that there exists a positive definite $W(x_1, x_2)$ such that $u > W$ for all x and t. If such a function exists then the origin is asymptotically stable. An interesting feature of this example is the fact that $\dot{V}$ is expressed in a time-dependent manner while V is not.

The previous examples considered only second-order systems. A network of nonlinear electrical elements provides a good example of a high-order system because the functions assume a particularly simple form if only inductors, capacitors, and resistors are present. It is shown in the second chapter that the state of an electrical network is described completely by the flux in each inductor plus the charge in each capacitor. Such a set of variables is called a *dynamic set* and it is the proper set of state variables for this method. As a Liapunov function it is convenient to take the energy stored in the reactive elements. This can be written as

$$V(i, v) = \sum_j \int_0^\infty i_j v_j \, dt$$

$$= \sum_j \int_0^{\phi_j} i_j \, d\phi_j + \sum_j \int_0^{q_j} v_j \, dq_j \qquad (7.45)$$

where the sum is formed over only the inductors and capacitors. If each inductor and capacitor has a positive sense such that $i_j\phi_j > 0$ and $v_jq_j > 0$ when the state variables are not zero then V is positive definite. The time derivative of V is given by the simple expression

$$\dot{V}(i, v) = \sum_j i_j v_j \qquad (7.46)$$

If each reactive element is connected to a passive resistive termination then $\dot{V}(0) = 0$ and $\dot{V} < 0$ for all i_j and v_j which are not zero. In this case the origin is asymptotically stable. If all resistors are replaced by short circuits then $\dot{V}(0) = \dot{V}(i, v) = 0$ and the origin is only stable. Under certain conditions a sustained oscillation may exist. The determination of conditions where Theorem 7.4 might apply is left as an exercise.

EXTENSIONS

Matrices are not used in this book because it is thought that their limited advantage in describing nonlinear networks and systems is outweighed by the necessity of introducing another branch of mathematics. However, almost all of the literature on the Liapunov method uses matrix notation and the reader who plans to pursue the subject further must prepare himself accordingly.

In addition to providing the ultimate in concise notation the use of matrices makes available a wide variety of results produced by many mathematicians working over many years. In particular, Sylvester has derived the requirements to form a positive definite function involving cross-product terms such as $a_{ij}x_ix_j$. The criterion states that all principal minors of the determinant of the a_{ij} must be positive. The proof of this can be found in books on advanced algebra. Expressions of the above type are very frequently used as Liapunov functions. Matrix methods are also useful in determining the properties of the linear part of the system and in calculating the time derivative of the Liapunov functions.

To establish asymptotic stability it is necessary first to establish stability so the two fundamental theorems by Liapunov essentially provide information about the performance of the system near the origin. From practical considerations it is desirable to extend this region. To do this it is first necessary to examine some topological properties of trajectories and to introduce some new mathematical concepts.

A set of points which contains only entire trajectories extending backward and forward forever is called an *invariant set*. Thus an equilibrium point is an invariant set and the set of an infinite number of trajectories is an invariant set.

If the system is autonomous then its trajectory does not wander aimlessly through phase space but it displays a definite tendency. If the trajectory approaches a fixed curve then the set of points on the curve is called a *limiting set*. An alternate definition of a limiting set is to say that given $\epsilon > 0$ there is a τ such that for all $t > \tau$ there is an x' in the limiting set such that $|x - x'| < \epsilon$. This means that the curve of the limiting set can be made as close as desired to the curve of a true trajectory. Assuming continuity in the behavior of the curves in phase space it is reasonable to assume that the limiting set is also a trajectory. This does not constitute an exact proof of the following lemma but hopefully it establishes its plausibility.

Lemma 7.1   *A bounded trajectory has a limiting set which is an invariant set.*

A limiting set is stationary in time so it must lie in regions where $\dot{x} = 0$. This can be related to a Liapunov function in the special case of an autonomous system wherein the relations

$$\dot{x} = X(x)$$

$$\dot{V} = \sum_i \frac{\partial V}{\partial x_i} \dot{x}_i \tag{7.47}$$

show that the set of points where $\dot{x} = 0$ lies in the set where $\dot{V} = 0$. (However, $\dot{V}$ may also be zero where $\dot{x} \neq 0$ so in general the set $\dot{x} = 0$ is a subset of the set $\dot{V} = 0$.) Therefore it can be concluded that a bounded trajectory approaches an invariant set lying in $\dot{V} = 0$. As an example, for the system

$$\dot{x}_1 = (x_1 - x_2)(1 - x_1{}^2 - x_2{}^2)$$
$$\dot{x}_2 = (x_1 + x_2)(1 - x_1{}^2 - x_2{}^2) \tag{7.48}$$

an appropriate Liapunov function is

$$V(x) = x_1^2 + x_2^2 \qquad (7.49)$$

Its time derivative is

$$\dot{V}(x) = 2(x_1^2 + x_2^2)(1 - x_1^2 - x_2^2) \qquad (7.50)$$

and $\dot{V} = 0$ at the origin and along the circle $x_1^2 + x_2^2 = 1$. Inside the circle $\dot{V} \geqslant 0$ so by Theorem 7.4 the origin is unstable. Since the trajectories cannot leave the circle they are bounded and they approach the invariant set at $x_1^2 + x_2^2 = 1$. The unit circle constitutes an invariant set because it is an infinite set of equilibrium points. It is important to observe that the circle itself is not a trajectory of the system.

The potential generality of the above principle can be demonstrated by observing the fact that in any given circle of radius R outside of the unit circle $0 < V \leqslant R$ and $\dot{V} < 0$ so the trajectories are also bounded and must again approach the invariant set on the unit circle. The following theorems from LaSalle and Lefschetz utilize these properties of bounded trajectories.

> **Theorem 7.6** If an autonomous system has a Liapunov function which is bounded on the exterior surface of a finite region then every trajectory starting in the region remains in the region and approaches an invariant set lying in the set of all points where $\dot{V} = 0$.

The Liapunov function has its greatest value on the exterior boundary of the region. Since $\dot{V} \leqslant 0$ in the region any trajectory starting inside must remain forever in it. Hence $x(t)$ is bounded and by the above principle it must approach an invariant set contained in $\dot{V} = 0$. As shown in the previous example there is no requirement for this set to be the origin. It can be anywhere in the region and the region can exclude the origin.

> **Theorem 7.7** If an autonomous system has a Liapunov function whose time derivative is negative definite and which is bounded on the surface of a finite region which includes the origin then every trajectory starting in the region approaches the origin.

This theorem which establishes asymptotic stability in a finite region is derived directly from the above stability theorem by observing the fact that the condition $\dot{V} < 0$ everywhere except the origin reduces the invariant set to the origin itself.

The previous theorems involve stability and asymptotic stability in a finite region. The next two theorems extend this region to the entire phase space. In this case the system is said to be *stable in the whole* or *asymptotically stable in the whole*. The greatest body of literature treats the latter case.

Theorem 7.8   *If an autonomous system has a Liapunov function which becomes infinite as $|x| \to \infty$ then all trajectories approach an invariant set lying in the set of all points where $\dot{V} = 0$.*

This theorem is derived directly from Theorem 7.6 by showing that the condition on $V(x)$ places a bound on every trajectory. This is easily done by observing that for any initial point x_0 and corresponding $V(x_0)$ there exists an R such that for all x where $|x| \geqslant R$ then $V(x) > V(x_0)$. Therefore since $\dot{V} \leqslant 0$ the trajectory can never leave the region contained in the hypersphere $|x| = R$. It is bounded and the theorem follows immediately.

Theorem 7.9   *If an autonomous system has a Liapunov function which becomes infinite as $|x| \to \infty$ and whose time derivative is negative definite then all trajectories approach the origin.*

In this case the trajectory is bounded and the invariant set consists of only the origin so the system is asymptotically stable in the whole.

Using a ratio of finite increments as an approximation for the time derivative $\dot{x}$ permits Equation 7.1 to be written as

$$x(t_{k+1}) = X[x(t_k), t_k] \, \Delta t + x(t_k) \tag{7.51}$$

which symbolizes a set of equations of the general form

$$x_i(t_{k+1}) = f_i[x(t_k), t_k] \tag{7.52}$$

Such an expression is called a *difference equation*. These equations arise in the analysis of *sampled-data systems* where the values of the state variables are known only periodically. The Liapunov method can be adapted to predict the nature of the solutions of difference equations.

A function $V(x, t_k)$ which is continuous in x, which satisfies $V(0, t_k) = 0$, which dominates a positive definite $W(x)$ for all $x \neq 0$ and all t_k, and whose difference

$$\Delta V(x, t_k) = \frac{V[x(t_{k+1}), t_{k+1}] - V[x(t_k), t_k]}{t_{k+1} - t_k} \tag{7.53}$$

is less than or equal to a negative semidefinite $W_2(x)$ for all $x \neq 0$ and all t_k is called a *discrete Liapunov function*.

Theorem 7.10   *If a system described by a difference equation has a discrete Liapunov function then the origin is stable.*

The proof is so similar to that of Theorem 7.1 that it is left as an exercise.

Theorem 7.11   *If a system described by a difference equation has a discrete Liapunov function which is dominated by another positive definite function and whose total time derivative is negative definite then the origin is asymptotically stable.*

The proof of this theorem is quite similar to that of Theorem 7.2. Again it is assumed that the motion terminates at a point other than the origin. The restrictions on $V(x, t_k)$ and $\Delta V(x, t_k)$ permit the selection of a $\mu > 0$ which is the minimum value of $|\Delta V|$. For an arbitrary time t_n the Liapunov function can be evaluated by

$$V(x, t_n) = V(x_0, t_0) + \sum_{k=0}^{n-1} \Delta V(x, t_k)(t_{k+1} - t_k)$$

$$\leqslant V(x_0, t_0) + \mu(t_n - t_0) \tag{7.54}$$

Since μ is a negative constant, for large enough n the value of V would become negative. This is impossible so the trajectory must terminate at the origin.

Other theorems such as the Chetaev theorem on instability can be adapted to discrete-time systems. This development is left as an exercise.

Before concluding this discussion it may be helpful to include an interesting idea from Kalman and Bertram. It applies to the general system described by Equation 7.1. In this technique one establishes the relations

$$\dot{V} = \left(\frac{\dot{V}}{V}\right)V \leqslant \eta V \tag{7.55}$$

where η is the minimum of the absolute value of the ratio of $\dot{V}$ to V over a region of the phase space where neither $\dot{V}$ nor V are zero. Integrating the above inequality gives

$$V(x, t) \leqslant V(x_0, t_0)e^{\eta(t-t_0)} \tag{7.56}$$

so η^{-1} can be interpreted as the maximum time constant in the region and η can be regarded as a figure of merit for the system. Unfortunately its value depends on the particular Liapunov function which is chosen so its interpretation must be regarded as only an approximation.

In conclusion it is probably wise to state once again that this chapter purports to be only an introduction to the Liapunov method. If the method appears to be applicable to a particular practical problem then it is recommended that comprehensive literature be consulted. At the present time a surprisingly large number of very general results have been obtained by the use of the method and though the solution of a particular problem may not be available it is quite likely that there are some good suggestions for solving the problem or perhaps even some related results which may be helpful.

REFERENCES

1. M. A. Liapunov, Problème général de la stabilité du mouvement, *Annals of Mathematics Studies*, Number 17, Princeton University Press, Princeton.

2. J. LaSalle and S. Lefschetz, *Stability by Liapunov's Direct Method With Applications*, Academic Press, New York, 1961.
3. W. Hahn, *Theory and Application of Liapunov's Direct Method*, Prentice-Hall, Englewood Cliffs, 1963.
4. R. E. Kalman and J. E. Bertram, Control System Analysis and Design Via the "Second Method" of Liapunov, *Transactions of the ASME, Journal of Basic Engineering*, 1960, pp. 371–400.
5. N. N. Krasovskii, *Stability of Motion*, Stanford University Press, Stanford, 1963.
6. A. M. Letov, *Stability in Nonlinear Control Systems*, Princeton University Press, Princeton, 1961.

INDEX